Jansenism

Catholic Resistance to Authority from the Reformation to the French Revolution

William Doyle

First published in Great Britain 2000 by
MACMILLAN PRESS LTD
Houndmills, Basingstoke, Hampshire RG21 6XS and London
Companies and representatives throughout the world

A catalogue record for this book is available from the British Library.

ISBN 0–333–68971–2 paperback

First published in the United States of America 2000 by
ST. MARTIN'S PRESS, INC.,
Scholarly and Reference Division,
175 Fifth Avenue, New York, N.Y. 10010

ISBN 0–312–22676–4 (paper)

Library of Congress Cataloging-in-Publication Data

p.   cm.
Includes bibliographical references and index.
ISBN 0–312–22676–4

This book is printed on paper suitable for recycling and made from fully managed and
sustained forest sources.

10   9   8   7   6   5   4   3   2   1
09   08   07   06   05   04   03   02   01   00

Printed in Malaysia

JANSENISM

Studies in European History

General Editor: Richard Overy
Editorial Consultants: John Breuilly
                        Roy Porter

In Memory of Desmond Charlton

Contents

Editor's Preface

The main purpose of this Macmillan series is to make available to teacher and student alike developments in a field of history that has become increasingly specialised with the sheer volume of new research and literature now produced. These studies are designed to present the state of the debate on important themes and episodes in European history since the sixteenth century, presented in a clear and critical way by someone who is closely concerned with the debate in question.

The studies are not intended to be read as extended bibliographical essays, though each will contain a detailed guide to further reading which will lead students and the general reader quickly to key publications. Each book carries its own interpretation and conclusions, while locating the discussion firmly in the centre of the current issues as historians see them. It is intended that the series will introduce students to historical approaches which are in some cases very new and which, in the normal course of things, would take many years to filter down into the textbooks and school histories. I hope it will demonstrate some of the excitement historians, like scientists, feel as they work away in the vanguard of their subject.

The format of the series conforms closely with that of the companion volumes of studies in economic and social history which has already established a major reputation since its inception in 1968. Both series have an important contribution to make in publicising what it is that historians are doing and in making history more open and accessible. It is vital for history to communicate if it is to survive.

R. J. OVERY

Preface

The unbelieving son of a lapsed Catholic, brought up as a 'non-playing Anglican', is perhaps not the obvious person to write about tangled episodes in the history of the Catholic Church. Nevertheless this book represents a long-standing interest in church history, and at least a partial fulfilment of an idea first conceived in the early 1970s. A proposal I then put to a Christian publishing house for a short book (based on lectures at the University of York) on eighteenth-century Jansenism was declined, as unlikely to sell. Yet in subsequent years this later period of Jansenism has become a major focus of historical interest, and unprecedented claims have been made for its importance in the origins of the modern world. And although most of this recent work has been focused on Jansenism's once-neglected later phases, new insights also continue to appear on its more famous 'classic' period. So the time now seems even riper for a fresh introduction to the whole subject, especially for English-speaking readers, who have always found it relatively inaccessible. In trying to distil it into the short space required by this series, I am only too aware of how much I have had to distort, or omit entirely. Many names and episodes, which specialists would regard as indispensable, find no mention here – among the absent names for example, are those of Barcos, Bossuet, Gerberon, Clément or Tamburini. I can only hope that the bibliography will lead interested readers to most of them. But in choosing what to leave in and out, I have been fortunate in the kind advice of two friends more learned in the history and doctrines of the Christian Church than I could ever hope to be. I am grateful to the Revd Professor Richard Bonney, and to Dom Aidan Bellenger, OSB, for reading and commenting on the entire text. To them, and to my wife who read it first, I am extremely grateful. The flaws they failed to save me from were doubtless predestined, and none of their responsibility.

Introduction: Jansenism and the Historians

Jansenism was the most persistent problem afflicting the Catholic Church for almost two centuries. And yet people known as Jansenists habitually denied that such a thing existed. What, then, have those using the term thought they meant?

The Jesuits who coined it in the 1640s had no doubts. Jansenists followed the doctrines and venerated the memory of Cornelius Jansen, a Dutch-born theologian who died as Bishop of Ypres in 1638. Jansen's doctrines, purportedly derived from St Augustine, proclaimed that God's children were powerless to affect their own salvation. God had already predestined them to be saved or damned. Such ideas, echoing as they did the main tenets of Calvinism, seemed more Protestant than Catholic. Accordingly they appeared heretical.

Those accused of heresy, however, denied it. The Truth was, they declared, that Jansen was no heretic, had left no disciples and had not propounded the doctrines ascribed to him. So-called Jansenism, they said, was an illusion foisted on the Catholic Church by a ruthlessly ambitious order anxious to suppress all criticism of their own theology of free will. Illusion or not, it was powerful enough to induce both ecclesiastical and lay authorities in France to persecute those they identified as Jansenists and the institution recognised as their spiritual headquarters, the female monastery of Port Royal. The history of Jansenism, down to the final obliteration of Port Royal's last vestiges in 1711, was largely the history of this persecution.

By that time, general histories of Jansenism were already being written; and in one sense the whole of the 'second Jansenism' of the eighteenth century was a historical debate about the first [63]. In attempting to define the heretical beliefs of Jansenists in the bull *Unigenitus* of 1713, the Pope found both his accuracy and authority disputed by new generations of self-styled 'Defenders of the Truth'.

1

In fact, much of the story of seventeenth-century Jansenism, as it is known to posterity, was first established by pious chroniclers in the century that followed. It was only after these disputes had in turn faded away – as the church confronted the challenge of the French Revolution – that doctrinal partisanship ceased to be the main driving force for those writing about Jansenism's history.

The French Revolution had not merely challenged the Church. An event so cataclysmic and unforeseen forced all who lived through and after it to think afresh about the history and identity of France. And Jansenism, which had flourished there as nowhere else and featured so prominently in the history of the old regime, inevitably began to be re-examined from this perspective. The importance of the question was enhanced by the literary and artistic eminence of some of the figures associated with Port Royal: the philosopher-scientist Blaise Pascal, the painter Philippe de Champaigne, the poet and playwright Jean Racine (himself an early historian of Port Royal). Historical and literary approaches came together in 1840–59 with the publication of *Port Royal* by the father of modern French literary criticism, C. A. Sainte-Beuve. This multi-volume epic, based on unprecedented original research, is still the indispensable starting point for all serious study of the subject. Sainte-Beuve's interest in the theological disputes of the seventeenth century was secondary to the human drama of his story. He was moved by the romance and tragedy of the heroic age of Jansenism; and he had no desire to pursue it beyond 1711. Although well aware of the second Jansenism of the eighteenth century, he saw it as 'shrivelled, dried-up, like a branch of a river running away into the sands and lost among stones ... allowing many to be of the party but not of the dogma, or even religious at all' [9, vi]. His attitude was shared by most of those who wrote after him. The true focus of the subject lay in the sublime days of the *grand siècle*. Post-*Unigenitus* Jansenism, more widespread though it was, was not about salvation. Its true concerns were political and legalistic, a mask for conflicts more material than spiritual. And in the form of the 'convulsionism' which occurred in its name in and after the 1730s, it was hysterical, degrading and almost impossible to make sense of.

These perceptions dominated the field until the mid-twentieth century. Only Augustin Gazier's general history of the Jansenist movement [6] gave as much weight to what happened after *Unigenitus* as before it; and Gazier's open commitment to exonerating his subject made him suspect to many scholars. How could over 600 pages seriously be devoted to a movement their author claimed did not really exist? Thus the most influential early twentieth-century work continued to concentrate on the great figures of the heroic age. While the historian of religious sentiment, Henri Brémond, laboured to reclaim Pascal and others for Catholic orthodoxy [1], three decades later the Marxist critic Lucien Goldmann attempted to depict Jansenism as the symptom of a social crisis [27]. Noting that many of the leading proponents of Jansenism's 'tragic vision' came from legal circles, Goldmann attributed their renunciation of worldly values to the decline of their class, the nobility of the robe. Seldom can a historical hypothesis have been more unanimously rejected by those who knew anything about the subject. Equally remarkable, however, was the range of reasons offered for rejecting Goldmann's ideas. They reflected the way that study of Jansenism had fragmented since the days of Gazier and Brémond. Jansenism now increasingly appeared less as a unitary movement than as a diverse range of tendencies within the Catholic Church, more easily identified by what they were not than what they were. But although these nuances were a tribute to scholarly rigour and discrimination, they scarcely made the subject more approachable, and the more accessible modern studies of seventeenth-century Jansenism have tended to ignore them [34, 45].

The most significant development of the later twentieth century has been a revival of interest in Jansenism after *Unigenitus*. Although pioneered as early as 1928, in Edmond Préclin's work on the origins of the revolutionary civil constitution of the clergy [68], it only flowered as historians became dissatisfied with socio-economic approaches to historical explanation. In those terms, what passed for Jansenism in the eighteenth century had indeed been well-nigh inexplicable. But a return to political history demanded some prior understanding of the Jansenist terms in which so many eighteenth-century political struggles were conducted. A new interest in broadly defined cultural history also opened the way at last to more sympathetic and sensitive analyses of the mass support which opposition to *Unigenitus* received, including the cult of convulsionism [59, 61]. Much of the new work came from English-speaking scholars, whose previous contributions to Jansenist studies, with two notable exceptions [4, 25], had been minimal. French interest in these later phases has, however, also begun to revive under the inspiration of Catherine Maire [7, 61, 62, 63]. The result has been to endow eighteenth-century

Jansenism with a historical importance if anything exceeding that of the earlier, heroic phase. Jansenists are now seen to have the prime responsibility for the expulsion from France of their traditional enemies, the Jesuits [78]. They are seen as the main source of discourses of resistance to royal authority articulated by the magistracy; and thus, ultimately, of the 'desacralisation' of monarchy which led to regicide, both before and during the French Revolution [64, 79, 81]. As a result, much that had always been intellectually attributed to the Enlightenment is now being reassigned to those whom the *philosophes* regarded as their direst enemies. And, in the process, it is becoming increasingly clear that underlying all that opponents of *Unigenitus* did was a religious commitment every bit as sincere and principled as that of Port Royal and its circle. The rediscovery of eighteenth-century Jansenism has laid bare all that bound it to the movement of the earlier century. At long last, the mould of interpretation cast by Sainte-Beuve seems to have been shattered. Whether that leaves Jansenism any easier to define is the main problem confronting this book.

1   The Seeds of Jansenism

When Adam ate of the forbidden fruit in the Garden of Eden, he defied the God who had created him. This was the original sin. Its consequences, borne by all Adam's descendants, were death, pain, human depravity and imperfection.

But God so loved mankind that He did not wish all Adam's seed to perish. He offered salvation from the consequences of sin by sending His only son into the world to sacrifice himself. The resurrection, when Christ triumphed over death itself, offered those who believed in Him the hope of the same redemption.

Not everyone, however, would choose Christian salvation – or attain it. And if God was almighty, all-knowing and eternal it was not obvious whether anyone had a free choice. An all-seeing God presumably knew in advance what choices every one of His creatures would make. An all-powerful God might be presumed even to have decided in advance what they would do. And in that case, Christ did not bring the offer of salvation to all of mankind. Some were predestined to accept Him and be saved, others to be damned; and human free will was at best an illusion.

These problems, and their implications, perplexed Christian thinkers from the earliest days of the Church. Late in the fourth century, Pelagius and his followers began to argue that faith and morality were meaningless unless the human will was genuinely free to choose or not to choose what is good. Pelagianism was condemned as heretical at the Council of Carthage in 418, largely thanks to the efforts of Augustine, Bishop of Hippo [14].

[i] Augustine

According to Augustine, fallen man is so depraved by the legacy of sin that he is incapable, unaided, of willing what is good; let alone

5

achieving salvation through faith in Christ. Divine assistance is essential, in the form of grace. And grace is a free gift of God. It cannot be earned, even by an abundance of good works, because God makes no bargains with His creatures.

Whether this left any scope at all for the free exercise of the will, or any incentive to live virtuously, remained problematical; although Augustine contended that there were senses in which it did. What is certain is that the defeat of the Pelagians established his overriding authority among the Church's early fathers. It is true that for a century after Augustine's death in 430 a small group of writers called Semi-Pelagians tried to reconcile the necessity of grace for achieving salvation with the freedom not to accept it. Their doctrines were included in a renewed condemnation of Pelagianism at the Council of Orange in 529; but by then such a condemnation amounted to little more than a restatement of orthodoxy. Nor did the authority of Augustine, the now-canonised 'Doctor of Grace', provoke further serious discussion for many centuries. But the efforts of the scholastics, from the twelfth century onwards, to apply reason and logic to the understanding of divine revelation led, by the fifteenth century, to a renewed emphasis in some learned circles on free will. There was talk of salvation being earned by spiritual effort and good works [19: 55–61].

Rigorists, notably among the many religious orders who followed a rule reputedly framed by St Augustine himself, sniffed Pelagianism. They denounced 'modern' backsliding into this old heresy with such sustained vigour that by the late fifteenth century a reinvigorated Augustinianism was sweeping Europe's centres of learning. It was the beginning of an 'Augustinian moment' [24: 6] that was to last for three centuries.

Its most spectacular manifestation was in the Reformation itself. What spurred Luther into breaking with the Church was an Augustinian realisation that good works are a result of the free gift of God's grace·rather than a means of earning it. And the driving force of Calvin's doctrines was an explicit belief in the predestination of the elect to salvation, and of the 'reprobate' to damnation. Augustine was his proudly acknowledged mentor. The appeal of the reformers to such an unimpeachable authority posed a serious problem for the Catholic Church when, after half-hearted attempts to stifle Protestantism, it formally recognised the need to reaffirm what it stood for.

[ii] Tridentine Catholicism

Although Protestants were intermittently present at the Council of Trent (1545–63) there was never much doubt that its purpose was not reconciliation but recovery. And to achieve that, two things were necessary. First, the Church had to make clear where, why and how it differed from the Protestants. Secondly, it needed to organise itself so as to win back those lost to heresy, convert the heathen, and maintain a firm hold on those who remained faithful. The complex, accident-prone and sometimes disreputable story of the Council's prolonged and much-interrupted deliberations, and the compromises and sudden changes of heart that lay behind many of its canons and decrees, should not distract attention from the clear orientation which it gave to the structure and doctrine of the Roman Catholic Church [17: *chs 1 & 2*].

Above all, it reaffirmed the authority of the Pope: so much so that no pontiff felt the need to convene another general council for 306 years. The Council submitted its decisions to papal confirmation and left several important issues to be finally decided in Rome. Although the outcome disappointed some bishops, hopeful of subjecting an overlord who had too often recently brought scandal on the Church to the control of his assembled peers, the Council also reaffirmed the authority of bishops over their flocks. Unlike many Protestant sects, the Roman Church would remain resolutely episcopal, with both clergy and laity firmly under their bishops' spiritual control and guidance. Among the highest episcopal duties would be to ensure that the parish clergy were better trained, and a target was set of a seminary in every diocese. But learning in the laity was viewed more ambivalently. The Bible and the liturgy would remain in Latin, inaccessible to individual interpretation by untutored minds. Though not authorised by the Council, it was also during its time that an *Index* of books prohibited to the faithful was first issued.

The fathers at Trent were well aware that Protestantism had begun when Luther took his doctrinal discontents beyond academic wrangling to appeal to a wider public. They were now determined to anathematise the errors that he and subsequent reformers had propagated. So there would be no priesthood of all believers; an ordained clergy, and a church to authorise its ministry, was necessary for conferring the sacraments which admitted the faithful to grace. And grace itself, and the 'justification' or righteousness which was its consequence, was not to be attained, as Luther had contended, by faith alone. Grace came also from the sacraments and from good works. All these acts were matters of choice. God's creatures were not passive recipients of His grace. Although God moved their will towards acceptance, they remained free to reject it. And although He had foreknowledge of all they would do, their fate was not predestined. How this could be, the Council did not choose to explain, but its reasons for emphasising the importance of the sacraments, good works and free will were obvious enough. They were, indeed, explicitly acknowledged by some of the participating prelates. Nobody would be converted, or reinforced in their faith, if they were told that nothing they did would make any difference to what God had already decided. The survival and propagation of the faith, and the continued vitality of the Church, required active Christians. They, in turn, needed incentives and the assurance that their own eternal salvation depended on what they freely chose to do.

Nobody believed this more fervently than the Jesuits. The Society of Jesus was not the only new order to be founded in the challenging decades which culminated with the Council of Trent, but it was incomparably the most successful. Its soldier founder, Ignatius Loyola, imbued it from the start with a military discipline and unquestioning obedience to the Pope. Not surprisingly, in 1540, within six years of vows taken by Ignatius and a small group of friends, the Society received formal papal recognition. Its main commitment was to the propagation of the faith, both by the conversion or reconversion of heathens and heretics, and by the better education of Catholic elites. These were matters in which the decisions of the Council were crucial, and one of the earliest successes of the Society was the influence which its determined theologians had on the final content of the canons and decrees.

Not everybody at Trent, however, welcomed this brash new force. Bishops were suspicious of an order which acknowledged no authority between its own hierarchy and the Pope himself [18: 133–6]. The Dominican order was also well represented at Trent. They were the traditional watchdogs and preachers of orthodoxy, and the Council, instinctively reaching back to the medieval scholasticism which the Protestants had spurned, reconfirmed the authority of the most illustrious of all Dominicans, St Thomas Aquinas. Like the Jesuits, the Dominicans recognised the dangers of over-emphasising predestination, but they were anxious to distance themselves from energetic

8

new rivals whose aspirations were an implicit reproach to their own failures. Finally, although their defeat tends to leave them overlooked, there were also at Trent a number of theologians who saw no reason to abandon the pure doctrines of St Augustine merely because the Protestants had appropriated them. One, indeed, had already alarmed the authorities even before the Council had finally refined the Church's official stand. He was Michel de Baye, called Baius (1513–89), sent to Trent as the elected representative of the university of Louvain.

[iii] University theology

Louvain was one of the leading centres of learning both in northern Europe and in the Habsburg domains. Proudly orthodox, it had been one of the first universities to condemn the works of Luther. It was also proudly modern: its professors were as determined as any of the reformers to retrieve pure texts of the scriptures and the early fathers, purged of medieval corruptions and glosses. Baius, who spent his entire scholarly life at Louvain, espoused this modernism in its most extreme form. He tried to read St Augustine as if centuries of scholastic interpretation had not intervened since the saint had first written. From intense study of the anti-Pelagian writings, Baius concluded that good works were impossible without the prior gift of grace [25: 81–92]. This seemed precisely Luther's position; and in the 1650s, as hopes faded of ever healing the Protestant schism, such similarities were increasingly regarded as dangerous. Baius's teaching was denounced. A number of his alleged propositions were laid (1560) before the Sorbonne, the most important theology faculty in Christendom – and condemned. Baius denied that the propositions represented his views, but tracts that he published between 1564 and 1566 provided enough evidence to bring him to the attention of Rome itself. In 1567, Pius V condemned 79 propositions attributed to Baius – although the bull *Ex Omnibus Afflictionibus* did not name him. Anonymity enabled him to deny that the bull had condemned any views of his; but after twelve years of prevarication the condemnation was renewed by Gregory XIII in the bull *Provisionis Nostrae* (1579). This time Baius formally recanted, but by then he had become a *cause célèbre* at Louvain and was notorious among Catholic theologians everywhere.

The prestige and influence of Louvain made it one of the first university cities where the Jesuits sought to establish themselves. Its growing theological notoriety persuaded them to send some of their most promising younger recruits there to hone their disputational skills against 'Baianism'. Among them, in 1585, was Leonhard Leys, or Lessius, already known at the age of 35 for his anti-Augustinian teaching. When, the next year, the general of the Jesuits encouraged the society's theologians to explore new doctrinal avenues, an underlying strategy became clear. The ageing Baius took up the challenge and in 1587 persuaded his university to condemn most of Lessius's known opinions. By the time Baius died two years later, the dispute had been referred upwards to Rome for judgement. This was because in 1588 the full significance of the Jesuits' initiatives had been made clear with the publication in Portugal of Luis de Molina's *De Concordia liberi arbitrii cum gratiae donis, divina praescientia, providentia, praedestinatione et reprobatione ad nonullos primae parte D. Thomae articulos.*

The very title was a challenge, evoking all the key issues of free will, grace, God's foreknowledge, predestination, and the authority of Aquinas, which lay at the heart of the century's theological quarrels. A Jesuit since 1553, Molina believed that there need be no contradiction between the necessity of God's grace and human freedom to earn it. God offered grace sufficient for salvation to all. But not all were bound to accept it, and only for those who did would it become efficacious and achieve its end. God knew in advance who would benefit and who would not, but he did not dictate or determine their choice, or the success or failure of their spiritual efforts. Divine foreknowledge was not predestination, but rather what Molina called God's 'middle knowledge' (*scientia media*), which was a 'most high and inscrutable comprehension of every free will'. [25: *109*] What that meant is scarcely clear; but Molina's purpose was obvious. It was to safeguard the freedom of mankind to earn salvation, rather than attain it in some divinely ordained lottery.

Doctrines tending to such a conclusion had been taught in Jesuit colleges for some years. Molina's innovation was to print and publish them. But in framing his book as a discussion of Aquinas (whose authority had been confirmed yet again as recently as 1567, when Pius V declared him a Doctor of the Church) he was challenging not just the disciples of St Augustine, but also the Dominicans. In the Iberian peninsula the Dominican order was as well-entrenched in theology faculties as the Baianists were at Louvain, and Molina's book was a salvo in a struggle for influence throughout the world of Catholic learning. The Dominicans tried to prevent its publication and subsequently made a determined effort to have it condemned by theology faculties throughout Philip II's dominions, not to mention the Inquisition. Inevitably such a serious conflict between leading expositors of the faith reached Rome. In 1594 the Jesuits were ordered to restate their acceptance of Aquinas's authority, but learned polemics continued and in 1597 Clement VIII established a special commission, the congregation *De auxiliis divinae gratiae*, to examine the issues. The Dominicans contended that 'middle knowledge' was a concept unknown to St Thomas and that the spirit of 'Molinism' was Pelagian. Although the Thomism of the Dominicans was far removed from the views of Baius, who had scorned all forms of scholasticism, the Jesuits argued that both were virtually Calvinistic. The congregation sat intermittently through the reigns of three popes, two of whom seemed at times on the verge of condemning Molinism. But rather than taint with heresy an order now inseparable from the revival of Catholic energies, in 1607 Paul V dissolved the congregation, forbade either side to accuse the other of heresy, and prohibited any further publications on the subject of grace. Unable to believe that this was really the end of the matter, members of both orders continued to prepare weighty volumes on the subject. In 1610, Lessius finally produced a vast treatise on efficacious grace. Once more the Louvain theologians denounced him, ignoring new papal attempts to impose silence in 1611.

Grace and predestination, in fact, remained the central theological issues they had been when the Reformation began. Even the Protestants were discussing them afresh. At the Synod of Dort in 1618–19, Dutch Calvinists found themselves defending predestinarian orthodoxy against what they saw as the Pelagian doctrines of the Arminians. The fact that Arminianism was perceived by many Protestants as crypto-Catholicism shows how far the Roman Church appeared to have abandoned the authority of St Augustine. Its refusal to allow any further discussion of grace reinforced the impression. This refusal was formally reiterated yet again in 1625. Yet, by an incredible oversight, this papal decree was never formally promulgated in the Spanish Netherlands, and so was never recognised by the university of Louvain.

2 The Birth of Jansenism

[i] Counter-Reformation in France

The Council of Trent succeeded in stifling any nascent Protestantism in Italy and Spain. In France it was too late. By the time the Council dispersed, religious civil war had already broken out and would dog French history down to 1629. These conflicts made French Catholics peculiarly sensitive to the problems of defending and strengthening the faith against heretics. When the Protestant Henry of Navarre became heir presumptive to the throne in 1584, a Catholic League was formed, with Papal and Spanish support, to resist his succession. Ironically it was a League-inspired fanatic who activated his claim by stabbing Henry III five years later. Another five years of even more frenzied resistance on the part of the League followed, and only came to an end after Henry IV's conversion to Catholicism in 1593 was accepted and recognised by the pope. Though genuine, so belated an apostasy did not allay the suspicions of many Catholics, and throughout his reign Henry was the object of repeated assassination attempts. In 1610 one of them eventually succeeded.

The legitimacy of murdering monarchs was one of the most sensitive issues raised by Europe's sectarian fragmentation [20: *86–105*]. Ever since 1555, when the religious wars of the Holy Roman Empire had reached a truce at the Peace of Augsburg on the principle of *cuius regio, eius religio* (to each kingdom, its own religion), it had been widely accepted that the religion of a prince and his subjects ought to be the same. In these circumstances, the salvation of millions might depend upon a ruler's personal convictions, and it is scarcely surprising that old doctrines of tyrannicide should receive renewed discussion. Monarchs were understandably alarmed, as were any subjects who valued the rule of law. But the killing of Henry III was applauded by some of the more fanatical adherents of the League as legitimate tyrannicide. Nobody could doubt that such people thought the heretic Henry IV even more deserving of the same fate. All supporters of the League were tainted by these associations, and none more than the French Jesuits.

Although Loyola and the founding fathers took their first vows there, the Jesuits had a bad start in France. Their unequivocal commitment to papal supremacy aroused deep suspicion. Only the onset of civil war in the 1560s had allowed their establishment in the teeth of opposition from the parlements, the sovereign courts who saw themselves as the guardians of the laws of the kingdom. At this early stage, Jesuit theological innovations attracted little notice. But the success of their schools and colleges made universities jealous, including the Sorbonne; and Jesuit support for the League made them seem like agents of the king of Spain. Their refusal to accept Henry IV, a king legitimate under French law, until his recognition and absolution by the pope, confirmed all the worst suspicions of jurists. As moderate Catholics, sickened by almost forty years of disorder, rallied to the new king, an attempt was made on his life in 1594 by an unreconciled Leaguer who had received a Jesuit education. The opportunity to blame the Society was irresistible. The Sorbonne demanded its expulsion from the kingdom, employing a rising star of the Parisian bar, Antoine Arnauld (1560–1619), to denounce it as an agent of foreign powers which condoned regicide [46: *ch. 2*]. The result was a concerted drive among the parlements in 1594–5 which brought the banishment of the Jesuits from most of French territory.

A few years later, suspicions were further fanned by the incautious equivocations of the Spanish Jesuit Mariana, who, in his book *De Rege et regis institutione* (1599), argued that in certain circumstances anyone might kill a tyrant. The taint of regicide would cling to the Society of Jesus throughout its subsequent history. Yet in France its banishment did not last long. The first generation of Jesuit pupils had now reached maturity, and positions of influence. The king was anxious to demonstrate his commitment to the Church, and gradually many of those with a past in the League came to accept him. In 1603 these influences reopened the whole of the kingdom to the Jesuits and within four years Henry IV had a Jesuit confessor. Despite the continuing protests of jurists, magistrates and many clergy, the Society was soon the dominant clerical influence at Court. During their brief existence, Jesuit theologians had carried to new heights the traditional science of casuistry. By seeking out diversity of opinion among patristic authorities, confessors could resolve penitents' moral and spiritual dilemmas by choosing among allowable courses of action. The resulting flexibility made Jesuit spiritual directors popular among the rich, the fashionable and the powerful in search of the least burdensome ways of salving their consciences. Accordingly, the Jesuits were too well entrenched among powerful supporters to be expelled again when Henry IV finally fell to an assassin's knife. Nor was their prestige seriously challenged until the 1640s, after the deaths of Richelieu and Louis XIII.

Yet beyond the world of the Court and high politics, many sincere Catholics remained suspicious of much that the Jesuits stood for. Their emphasis on the authority of the Pope ran against the deep-rooted tradition of Gallicanism. Entrenched among civil and canon jurists, and reinforced by the alarming extremism unleashed by the League, Gallicanism denied the pope's jurisdiction over the French Church. Some Gallicans even questioned Rome's spiritual authority, preferring that of councils. Even then, many of the canons and decrees of the Council of Trent were never recognised by the courts as legally binding in France; and some Gallicans took the view that within the kingdom there was no higher ecclesiastical authority than a council of the French Church. The leading Gallican theorist of Henry IV's reign went further still. Edmond Richer (1559–1631) elected syndic of the Sorbonne in 1608, published his *De ecclesiastica et politica postestate* in 1611. While emphasising that the only infallible authority in the Church was a council, he argued that what gave councils authority was election by the entire body of the priesthood. The basic function of the Church was cure of souls. Accordingly, the parish priests who fulfilled that responsibility, and enjoyed spiritual powers no less extensive than any member of the hierarchy, should have the decisive voice in its affairs. They should meet regularly in synods and elect their bishops. The episcopate, and the pope himself, were mere executives with delegated powers [37: 72–5]. Such views naturally alarmed the hierarchy, not to mention regulars without cure of souls; and in 1612 Richer was deposed as syndic on governmental instructions.

[ii] Bérulle

Within the Sorbonne, however, Richer remained influential until his death; and he blamed his deposition on the machinations of rival advocates of priestly power in the newly established congregation of the Oratory [33: *106–16*]. Inspired by looser Italian models of half a century earlier, the French Oratory of Jesus and Mary was founded in 1611 by Pierre de Bérulle (1575–1629). Its purpose was to promote the spiritual efforts of ordinary priests. Oratorians lived in endowed communities without cure of souls, but took no vows. Their commitment was to serve the laity through administering the sacraments, improving the training of priests and educating the young. Unlike regular orders, the Oratorians unambiguously acknowledged the authority of bishops. Bérulle had been a pupil of the Jesuits and he admired their achievements, their discipline and their Christ-centred devotions. But he believed that comparable spiritual influence could be attained without vows, through the proper exercise of priestly powers alone. The idea was a great success. By the time of Bérulle's death there were 73 Oratorian communities throughout France, four seminaries, and 17 colleges. Oratorian educational ambitions soon alarmed the Jesuits, and they made some clumsy efforts to embarrass Bérulle, which sowed the seeds of a long and bitter rivalry. But between Henry IV's death and his own, Bérulle was too influential to be easily undermined. His talents were as much in demand in the political world as in the spiritual, and his aspiration was to bring the two ever closer together. The guiding impetus of state policy, he believed, should be to promote the interests of the faith; and these views were shared by an important group at Court who became known during these years as the devout (*dévots*). With roots in the League, they favoured an uncompromising approach to all Protestants. They were able, in 1627, to secure Bérulle a cardinal's hat. But when, after the final destruction of Protestant military power in France in 1629, they sought French involvement on the Catholic side in the Thirty Years' War, they came into fatal conflict with the cardinal minister Richelieu, who put the state's interests before those of the faith. Bérulle died before the crisis of 1630 which brought the rout of the *dévots* and secured Richelieu permanently in power. But the Oratory continued to flourish, and those touched by Bérulle's influence remained powerfully entrenched throughout the French Church.

Bérulle reached his convictions about the primacy of the secular priesthood after a decade as spiritual director to a number of pious ladies. The consolidation of the Catholic establishment after Henry IV's conversion favoured a remarkable efflorescence of female spirituality [22]. New orders of women were established, exemplary foreign ones introduced, and existing foundations reformed. Directly or indirectly, Bérulle was involved in most of these initiatives: but one at least had nothing to do with him. The initial reform of the Cistercian convent of Port Royal was the sole work, it seems, of its young abbess, Jacqueline Arnauld, better known by her monastic name of Mère Angélique (1591–1661). The fourth child of that same Antoine Arnauld who had denounced the Jesuits in 1594, she was thrust unwillingly into monastic life at the age of nine as abbess-designate of this run-down house not far from Versailles. Although abbess from 1602, it was not until six years later, as a result of a conversion experience during a sermon by an itinerant preacher, that she found her monastic vocation. The convent was now subjected by her to the strict primitive rule of St Benedict, and in 1609 she introduced enclosure, excluding even her family from the precincts (*journée du guichet*, 25 September 1609) [46: *ch. 3*]. Her reputation for austerity soon boosted recruitment and in 1626 the community of Port Royal moved to bigger premises in the southern suburbs of Paris. There, under *dévot* patronage, it was allowed to secede from the Cistercian order and take on a separate identity under the name of the Institute of the Holy Sacrament. Like all female communities, Port Royal needed a spiritual director in Holy Orders. Between 1625 and 1633 this was Sébastien Zamet (1587–1655) bishop of Langres. A disciple of Bérulle and well-connected at Court, Zamet was an influential protector of Angélique's aspirations. Thus, when a pamphlet defending the new order by her sister, Mère Agnès (1593–1672), was denounced for heterodox tendencies, Zamet found a tried theological polemicist to write in her defence. Subsequently he succeeded the bishop as Port Royal's spiritual director. The polemicist was Jean Duvergier de Hauranne, abbot of Saint-Cyran (1581–1643) [30].

[iii] Saint-Cyran

From a rich, devout Basque family, Saint-Cyran was educated by the Jesuits, completing his studies with them at Louvain between 1600 and 1604. He was destined from his earliest years for the Church, became a priest in 1618, and was soon afterwards provided to the benefice by whose name he is remembered. He owed this appointment to the bishop of Poitiers, a fighting prelate whose armed exploits against local Protestants Saint-Cyran had defended in a controversial pamphlet. He had made his reputation first as pamphleteer in 1608 when, with brilliant casuistry, he had argued that suicide was not only legitimate, but a duty, if it saved the life of one's sovereign. The moral relativism and subtlety of argument in these writings bore witness to the thoroughness of a Jesuit education; they seem evidence, too, of worldly ambition. But there was another side to Saint-Cyran which found expression in the friendship he made in Paris in 1609 with a fellow graduate of Louvain (though not its Jesuit college), Cornelius Jansen (1585–1638). They shared a passionate interest in the scriptures and the Church's early fathers. They spent the years 1611–14 studying them together in retreat on Saint-Cyran's family estates; and they corresponded regularly, and met periodically, over the next twenty years. Saint-Cyran's loyalty to his learned friend, and later to his memory, was the spark that ignited Jansenism and gave it a name.

Meanwhile they went their separate ways: Jansen back to Louvain, where he was ordained and taught theology, Saint-Cyran to Paris and the fashionable world of religious politics. At the end of the decade, however, the lives of both men were to change direction decisively. Jansen discovered St Augustine and Saint-Cyran met Bérulle.

At Louvain, where the memory of Baius was still revered and Lessius still taught, it was impossible to forget the vexed question of grace. A fresh attempt by the Jesuits in 1619 to penetrate the university provoked renewed efforts to expose the dangers in their doctrines. Jansen followed them in detail. He was also intrigued, as a Dutch speaker, by the debates at the Synod of Dort. The Dutch Calvinists, he observed to Saint-Cyran, 'follow almost entirely the doctrine of Catholics in the matter of predestination and reprobation' [28: 53]. From this time on, he began to read St Augustine avidly. 'I love him uniquely,' he wrote a year later [29: 65]. 'It seems to me that nothing among the ancients and moderns comes near to him by a hundred leagues. And the more I read him, the finer I find him.' During 1621 his enthusiasm began to crystallise into a plan to write a definitive commentary on the saint; stripping him, as Baius had tried to do, of centuries of misleading glosses. Jansen knew how dangerous this was. Even in his letters to Saint-Cyran he only spoke of his project in code, calling it *Pilmot*. And it progressed slowly, taking until 1638 to complete. By then Jansen was close to death, and he had to entrust publication to his will's executors. No inkling of his obsession had appeared in his teaching over the years. If he had been suspected

17

of unorthodoxy he would certainly not have been made bishop of Ypres, as he was in 1636. This promotion, indeed, came as a direct reward for an act of Catholic zeal: in a pamphlet entitled *Mars Gallicus* (1635) Jansen denounced French intervention on the side of the Protestant powers in the Thirty Years' War. In Louvain, however, he was known as an enemy of the Jesuits, having travelled all the way to Madrid in 1625 to denounce their academic ambitions. In the intimate, intemperate little town the Jesuits of the college soon learned that a posthumous treatise by their old adversary was being printed. For two years they made resourceful attempts to prevent its publication on the grounds that it discussed the forbidden subject of grace. But because the papal prohibition of 1625 (see p. 11) had never passed into local law their efforts were unsuccessful. Jansen's vast Latin treatise, called *Augustinus*, was published in August 1640.

By then Saint-Cyran was in prison, arrested in 1638 on the orders of Richelieu. He had known Richelieu since they had been young clerical careerists together at Court. But once Saint-Cyran encountered Bérulle in 1620, their paths diverged. By 1622 Richelieu was a cardinal, and two years later a minister. Saint-Cyran, meanwhile, had renounced secular ambition. The ubiquitous Bérulle had produced a conversion in the newly ordained and hitherto very worldly priest. He admired his energy and the range of his patristic learning. Saint-Cyran in turn was overwhelmed by Bérulle's combination of mystical spirituality and practical determination to promote the faith. They spent hours together initially, and kept in regular contact until Bérulle's death. Afterwards Saint-Cyran remained devoted to his memory. He spent the 1620s promoting causes close to his new mentor's heart. In 1623 he enlisted Jansen's support for introducing the Oratorians into the Spanish Netherlands. He sought prestigious endorsements for Bérulle's writings and ghost-wrote replies to his growing number of critics. Inevitably these activities brought him into conflict with the Jesuits, whose flexible doctrines Bérulle increasingly deplored, as much as they in turn resented the growing influence of his Oratorians. These antagonisms culminated in 1626 with an all-out attack by Saint-Cyran on the 'faults and falsehoods' of a fashionable Jesuit author, whose reputation he destroyed. The victory was easy: but in achieving it Saint-Cyran was finally convinced of how insidiously dangerous his former teachers were. And so when, in 1632, English Jesuits operating out of France challenged episcopal authority over their activities, he saw it as an opportunity to attack the Society in more general terms. Written anonymously at first, *Petrus Aurelius* was a pamphlet series which denounced every aspect of Jesuit activity. It was also a celebration of the powers of bishops and the superiority of secular priests. It was this which persuaded Zamet that Saint-Cyran would be an effective and reliable defender of Port Royal.

The Arnaulds had known him since the early 1620s, when Arnauld d'Andilly (1589–1674), the most politically active son of the great advocate, introduced Bérulle's new disciple to his sisters [46: *ch. 6*]. At that time the main spiritual influence over Mère Angélique was François de Sales, the evangelising bishop of Geneva, later canonised in 1665. But he died in 1622, and for over a decade Angélique looked in vain for a suitable successor. Zamet was more effective as a patron and protector than as a pastor. Her search ended with Saint-Cyran. The Port Royal nuns loved his austere and uncompromising spiritual direction, rooted in a sense of bottomless human depravity. Sins could only be forgiven, and grace attained, if penitents' love of God made them sincerely contrite. Attrition – penitence owing to mere fear of eternal punishment – was not enough. Until he was satisfied of unfeigned contrition, the confessor would forbid communion, the ultimate sacrament. The spiritual efforts required to meet such exacting standards were all-absorbing. Perhaps they were only possible to those who abandoned the world and its temptations entirely. This was certainly the view of Antoine Le Maître (1608–58), a brilliant young lawyer related to the Arnaulds, who in 1637 gave up a promising career in royal service for a life of contemplative solitude near Port Royal.

Soon imitated by others, Le Maître's gesture created a sensation; and it was known to have been made on the advice of Saint-Cyran. By now the latter's enemies were multiplying fast. With the defeat of the *dévots* in 1630 he had already lost his most influential protectors. And whereas Jesuit enmity incurred by *Petrus Aurelius* was offset by a warm welcome from bishops, Saint-Cyran's association with Port Royal brought him no credit with Richelieu, who suspected it to be a nest of heretics from at least 1633 [25: *177–8*]. As bishop of Luçon, Richelieu had commended the doctrine of attrition; and he found frequent communion after formal penance a convenient salve to his own conscience as a practical statesman. Saint-Cyran's contritionism he thought too demanding, and dangerous too if it deprived the king of valuable servants like Le Maître. The cardinal's close confidant, the Franciscan Father Joseph, resented the way Saint-Cyran's influence had supplanted his own in a fashionable convent under his patronage. Zamet soon found the same happening at Port Royal, and complained bitterly. Meanwhile, in 1635, on the outbreak of the war against the Habsburgs so long feared and condemned by Saint-Cyran's old *dévot* patrons, his friend Jansen published *Mars Gallicus*. It was soon circulating in French, to the fury of Richelieu, who knew of their friendship. The last straw for the cardinal came in 1637 when, for complex political reasons, he decided to seek an annulment of the marriage of the king's turbulent brother, Gaston d'Orléans. For this, he needed support from theologians. Saint-Cyran was too well known not to have his approbation sought; but he refused it, even when tempted with the bishopric of his native Bayonne. Richelieu knew that in 1630, only days before the *dévots* were overthrown, Saint-Cyran had been ready to accept this same see from them. Refusal now was therefore an act of personal defiance on the part of one who seemed to be increasingly the rallying point for all who opposed the cardinal. It was time to silence him. The occasion came in the spring of 1638, when Saint-Cyran was accused of inspiring, if not actually writing, uncompromising commentaries on a new translation of St Augustine's *Of Holy Virginity*. If Luther and Calvin had been shut up as soon as they began to 'dogmatise', Richelieu observed, much trouble would have been avoided. But the same might be said in reverse of the action he now took. On 14 May 1638 Saint-Cyran was arrested. He remained imprisoned at Vincennes until after the cardinal died, and only survived his release by a few months, dying in October 1643.

Richelieu had created a martyr. Henceforth, Saint-Cyran would become an object of veneration to those who had been his friends – and of suspicion to the many elements he had antagonised. The first test of these alignments came in 1640, with the publication of his old friend's posthumous treatise on St Augustine. And this meant that what probably deserves to be called Cyranism [24: *183*] came together instead around the name of Jansen.

3   Right and Fact, 1640–69

[i] *Augustinus*

It took Jansen three Latin volumes and 1300 double-columned pages to expound the thought of the Doctor of Grace – stripped (as he saw it) of subsequent deforming commentary [25: *126–53*]. Much of his text was a historical survey of the Pelagian and semi-Pelagian heresies, and how Augustine had provided the key to rebutting both. But there was no pretence of detachment. Jansen wrote to secure Augustine's triumph in Catholic doctrine. Salvation, he set out to prove, would not be achieved without the impulsion of God's grace; and once grace was imparted it was irresistible. Even before the fall, Adam had needed a *sufficient* grace in order to desire what was good. After the fall, his sinful descendants needed more. Only *efficacious* grace could prevent them from sinning; and no effort of their will could guarantee that such grace would be granted. God had already chosen its recipients in advance; and those not chosen were predestined to damnation.

Little of this was new, as Jansen would have been the first to claim. His whole purpose was to combat innovation, which risked resurrecting the heresies that Augustine had exposed. Like Baius before him (whose reputation he was determined to rescue) Jansen blamed the corruption of true doctrine on rash attempts since the later Middle Ages to clarify the inscrutable will of God by the light of all-too-fallible human reason. Nor did he hesitate to name the most recent exemplars of this tendency. They were mostly Jesuits, too concerned to preserve some element of free will in the attainment of salvation. It was this prolongation of an old Louvain quarrel, on a topic where successive popes had tried to impose a smothering silence, that initially made *Augustinus* controversial. But the Jesuits, celebrating their first hundred years in 1640 with much self-congratulation, were in no mood to let attacks on them pass unanswered. Having failed to prevent printing and publication with legal arguments, they now sought ways of suppressing Jansen's book on doctrinal grounds. When they did, it was almost inevitable that their other enemies should rush to Jansen's defence.

Among them was Saint-Cyran, who had closely followed the progress of his friend's writing over the years. After his first few months in prison, during which officials sent by Richelieu to interrogate him failed to detect any credible evidence of heresy, he was allowed free contact with the outside world. Thus he soon learned of *Augustinus*'s appearance, and made efforts to promote a French edition. Meanwhile he continued his work of spiritual direction. His inspiration had always been Augustinian, in its emphasis on the depths of human depravity to be redeemed. The profound sinfulness of the world was what had led the 'solitaries' whom he inspired to shun it. The difficulty of shaking off sin was what made him a contritionist director, advising deferral of communion until true sorrow had been established. Jansen now provided a powerful theological underpinning for all this, a comprehensive challenge to everything that the Jesuits stood for: a theology of human freedom, a Christian engagement with the world, regular communion and the cumulative spiritual benefits of attrition. Anything new by the author of *Mars Gallicus* and correspondent of Saint-Cyran was also a challenge to Richelieu, and when he died the cardinal had begun to orchestrate an attack on *Augustinus* in France. The Jesuits of Louvain had already denounced it in 1642 as reviving the heresies of Calvin and Baius. Saint-Cyran did not live to undertake Jansen's defence; but before his death he found someone who would.

[ii] Arnauld

Antoine Arnauld (1612–94) was the youngest son of the Jesuits' old adversary and brother of the reforming abbess of Port Royal. Destined at his mother's urging for a clerical career, he met Saint-Cyran in his formative years and was introduced by him to Augustinianism in its most uncompromising form. By the time he was ordained priest in 1641 he had undergone a conversion similar to that of his sister in 1608, or Saint-Cyran himself on meeting Bérulle, and placed himself under the spiritual direction of the prisoner of Vincennes. Adjured by his dying mother, herself by then a nun at Port Royal, to dedicate his life to defence of the Truth, he had already agreed before Saint-Cyran died to undertake Jansen's vindication. But his first publication appeared just before his mentor's death, and was written under his supervision [46: *ch. 7*].

Of Frequent Communion was a defence of Cyranist penitential doctrine. Provoked by the case of a fashionable courtier going to a ball straight after communion, it argued that over-familiarity with the eucharist debased its importance and allowed penitents no time to consider its awesome significance. It was a contritionist manifesto, and written not in clerical Latin like the *Augustinus*, but in French accessible to anyone who was moderately literate. Saint-Cyran had also written pamphlets in the vernacular, but not on the central issues of his thinking. Arnauld now set these out, and his name alone was enough to win an audience for his denunciation of 'that deplorable abuse of imperfect confessions, of overhasty absolutions, of vain satisfactions and sacrilegious communions' [quoted in 9: ii, *177*]. But Arnauld also invoked the authority of the Council of Trent and that universal hero of the Counter-Reformation, archbishop Charles Borromeo of Milan (1538–84; canonised 1610). His book was endorsed by many bishops and theologians; and even its critics recognised its success by complaining that, in the years following its publication, communions in the fashionable churches of Paris fell away dramatically. As so often, attacks on the book made people read it; and the fury of intemperate Jesuits who denounced Arnauld from the pulpit as a scorpion, a serpent and a lunatic rebounded against them. When it was suggested, with the approval of the new chief minister Mazarin, that Arnauld should go to Rome to explain his views, both the magistrates of the parlement and Arnauld's fellow doctors at the Sorbonne protested at such an affront to the liberties of the Gallican church. Jansenism may have originated at Louvain, but what enabled it to spread so effectively was the unique protection it would receive from the anti-papal traditions of French law and the French Church.

Arnauld did not go to Rome. By now, indeed, he had committed himself to positions that would have made it extremely unwise to do so. In 1644 he fulfilled Saint-Cyran's last wish with the publication of an *Apologia for Jansenius* which expressed the ideas of *Augustinus* for the first time in French [25: *212–15*]. Although anonymous, its authorship was an open secret. Moreover, the work he was defending had now been formally condemned by the pope. After some vicissitudes the Louvain Jesuits' denunciation of *Augustinus* in 1640 had reached

Rome, where it provoked an examination by the Inquisition. The result was the bull *In Eminenti* (1643), which renewed the prohibitions on discussing grace issued after the congregation *De Auxiliis*, and condemned Jansen for reviving the errors of Baius. A mass of typographical mistakes and irregularities raised doubts about the bull's authenticity, which Arnauld was quick to exploit. The parlement and the Sorbonne also refused initially to accept it. In any case, nobody in France took it seriously, except as a further pretext for attacking Arnauld.

Arnauld was now positively inviting such attacks. His criticism of the Jesuits had not remained implicit. As early as 1643 he had denounced the technique for which they were most notorious – their casuistry – in his *Moral Theology of the Jesuits*. They replied by calling him a crypto-Calvinist, dedicated to the subversion of the entire Church. By the mid-1640s, the term 'Jansenist Party' was in general use to describe Arnauld and his circle, although they never used it themselves. In their own eyes they were simply 'Friends of the Truth'. And indeed they were still more of a family than a party: without the Arnaulds and their extensive network of relatives and intimates, the early history of Jansenism is inconceivable. But their solidarity was impressive, and in Port Royal they seemed to have a physical as well as a spiritual headquarters. In the 1640s, a new church was built for the Paris house, and the original buildings 'in the Fields' were reoccupied by Angélique and a handful of chosen sisters. By then, following Le Maître's example of renouncing the world, a colony of distinguished male 'solitaries' had refurbished the site; and they had also begun to offer elementary education to the children of sympathisers in what became known as the 'Little Schools'. The number of solitaries fluctuated: but there were never more than a dozen at any time. Nor were there ever many pupils in the Little Schools (although in the 1650s they would include Racine). To a suspicious eye, however, all this looked like more than mere pious eccentricity. It appeared to be an attempt to establish, and perpetuate, a sect subscribing to the heretical doctrines of Jansen and Saint-Cyran. What was worse, sympathy for these doctrines seemed to be gaining ground. Port Royal never seemed to lack powerful patrons and protectors at Court. And so, although Arnauld's first burst of polemical zeal had died down by 1645, frenzied attacks on him and his associates continued. Usually in French, this virulence kept public interest alive and probably did as much to arouse sympathy for Jansenism as to refute it. That was certainly the effect of a carefully premeditated manœuvre that came to fruition in 1649. In July that year, the Sorbonne was asked to consider a series of propositions which, if condemned, would entail a condemnation of Arnauld and his circle, too.

[iii] The five propositions

None of these controversies occurred in a political vacuum. Without the deaths of Richelieu and Louis XIII they might not have taken place at all. What allowed them to flourish was the weak and distracted state of a regency government at war, whose resources were running out. By 1648 the crown was facing bankruptcy and the defiance of the kingdom's leading office-holders. This confrontation inaugurated five years of upheaval and civil war, known as the Fronde.

Jansenists had little to do with provoking this series of conflicts, and the leading custodians of the memory of Jansen and Saint-Cyran behaved discreetly throughout this period. A once-influential argument that early Jansenism reflected the discontents of the robe nobility of office-holders now commands little support, and it is hard to imagine why it ever had more [27, 28]. But those now being called Jansenists had a reputation for independence which made the queen-regent and her Italian favourite Mazarin instinctively suspicious. Arnauld's brother d'Andilly, Jansenism's chief spokesman in courtly circles, was closer to the *frondeurs* than to Mazarin: and a long history of protection by Gondi, archbishop of Paris, and his nephew and co-adjutor, did Port Royal and its circle no good when the latter (who became cardinal de Retz in 1652) put himself at the head of the Parisian Fronde. There was thus no inclination on the government's part to stifle the new attack on *Augustinus*.

For although it was denied that the dubious propositions, reputedly collected from a number of recent bachelors' theses, had anything to do with Jansen, nobody was deceived. Three of the propositions mentioned grace, another free will, two others the semi-Pelagian heresies (see Appendix). When the faculty agreed to set up a commission to investigate them, 61 doctors denounced the action to the parlement. Arnauld, a Sorbonne doctor himself, rushed into print to defend St Augustine. The parlement attempted to impose a truce, but the Sorbonne was soon implacably divided, and the noise of the controversy gave orthodox bishops the opportunity they were looking for. Animated by Habert, the first cleric to have denounced *Augustinus* from the pulpit in 1642, and now bishop of Vabres, 85 of them signed a letter to the pope asking him to clarify doctrinal confusions that were disturbing the peace of the French Church. A dozen other bishops wrote to protest; but in 1651 the regent herself urged Rome to settle these disputes by condemning the now-notorious propositions. The result, in 1653, was a new bull, *Cum Occasione*, which condemned five of the disputed propositions (there had originally been seven) – four as heretical and the fifth as false. No mention was made of St Augustine, but the issue was declared to have arisen from the printing of Jansen's *Augustinus*.

This was far less equivocal than the bull *In Eminenti*. Innocent X would probably have preferred to be less definite, but by 1653 he felt obliged to defer to the French government. The Fronde had at last begun to crumble and Mazarin was back in control. The arrest of cardinal de Retz in December 1652 showed the favourite's renewed power. However, attempts to deprive the rebel cardinal of his archbishopric (inherited in 1654) rebounded on Mazarin – especially since by then Retz had escaped. The parish clergy of Paris objected loudly to the persecution of their rightful superior by an Italian favourite. They orchestrated resistance in regular assemblies until Retz's negotiated resignation in 1662: effectively the last Fronde, a religious one [36]. Faced with the possibility of an uncongenial superior being imposed on them, these assemblies espoused doctrines of priestly power first elaborated by Richer 40 years earlier [see above, p. 14]. They also gave vent to an outburst of frustration against persistent interference by the Jesuits in parish affairs, which led the dissident priests to denounce the laxity of Jesuit moral teaching and confessional practices in terms which no 'Friend of the Truth' could disagree with. By the late 1650s, indeed, leading Jansenist writers were producing pamphlets for the dissidents and 38 out of the 68 holders of parish benefices in the capital between 1653 and 1662 declared their sympathy at one time or another for Jansenist doctrines [37: *143–51*]. Clearly by the 1650s Jansenism was spreading beyond the Port Royal circle and was being taken up by other rebels against authority in the Church.

No wonder, then, that on receipt of *Cum Occasione*, Mazarin moved quickly to secure its acceptance by the French Church through an assembly of bishops. Jansenists, though disappointed, appeared to have felt that they must accept the bull as it stood. But others were equally disappointed that it had not been even more explicit in attributing the condemned propositions to Jansen, and the bishops were persuaded to petition Rome for a clarification. In 1654 the pope provided one. The five propositions, he declared, were Jansen's, and a true and accurate expression of his meaning. Those who defended him risked heresy; the letter announced that the papal Inquisition had condemned a range of works upholding or defending Jansen's claims to orthodoxy. This went even further than the bishops had hoped. Indeed, it complicated their strategy since the parlement of Paris refused to recognise the Inquisition's authority in France. Once again juristic Gallicanism lent the Jansenists crucial room for manœuvre. Nevertheless, the hierarchy now had the unequivocal papal lead it had sought, and in 1655 a group of bishops convened by Mazarin declared that all clergy, both secular and regular, should subscribe to a declaration or 'formulary' abhorring the five propositions, acknowledging that they came from Jansen's *Augustinus* and represented a wrong interpretation of the saint's doctrine. It was a signal for all-out attack; and although many bishops ignored the call in the interests of harmony within their dioceses, zealous lower clergy sometimes felt entitled to take action of their own. Among them was the confessor of the duke de Liancourt, a peer long associated with the Port Royal circle. Two weeks after the episcopal assembly, he refused the duke absolution unless he promised to sever his links with such a nest of heretics. Not for the last time (see below, pp. 60–1) a refusal of sacraments to suspected Jansenists rekindled controversies that might otherwise have died away.

Certainly it revived Arnauld's polemical energies. In his *Letter from a Sorbonne Doctor to a Person of Condition* he claimed that no ordinary priest could impose what was in effect an excommunication. Several months later, his *Second Letter to a Duke and Peer* went further. It reiterated most of the doctrinal stands he had taken up since 1643, with a capital addition concerning the now-notorious five propositions. Arnauld publicly accepted their condemnation; but he refused to accept that they were Jansen's, because they were not to be found in *Augustinus*. The pope had the right to define doctrine, but he could not define facts. The propositions were heretical, but they were not in the book.

In a narrow textual sense he was correct. Generations of scholars have combed the dense columns of the *Augustinus* without finding the exact wording used in *Cum Occasione* [25: *appendices i–iii*]. But phrases very close to it certainly can be found, and in the context of Jansen's overall arguments none of them substantially distorted his clear meaning. Anything else would have made nonsense of his whole enterprise. Arnauld must have known this, and later he more or less conceded as much. But by then it was too late. The attempt to save Jansen from condemnation on a technicality had offered his enemies yet another opportunity, and in November 1655 his two *Letters* were denounced to the Sorbonne. The government, exasperated by his defiant hair-splitting, insisted that the faculty condemn him. In February 1656 it duly voted for his expulsion and soon afterwards that of all other doctors who had voted in his favour.

Apart from the imprisonment of Saint-Cyran before Jansenism had even acquired its name, the expulsion of Arnauld was the first act of physical persecution against the Jansenists. Until now any attacks had been polemical, doctrinal, or in the case of refusal of sacraments, spiritual. Mazarin and the forces of orthodoxy were determined to pursue them now on all these fronts; but they were held up for much of the year 1656 by the anti-French inclinations of a new pope. Alexander VII did not like the Jansenists. He was the first to suggest that their spiritual headquarters at Port Royal should be closed down. But he also distrusted Mazarin, and tried to temporise when pressed to intervene afresh. The Friends of the Truth, however, did not wait for further pontifical fulminations; and they were now reinforced by new voices from a generation too young to have known Saint-Cyran.

[iv] Pascal and Nicole

Blaise Pascal (1623–62) was one of the towering figures of the seventeenth-century intellect, responsible for major advance in physics and mathematics, and providing in his *Pensées* (posthumously assembled and published by Jansenists) a powerful series of arguments for the Christian faith which have transcended denominations. But in the history of Jansenism he was above all a polemicist. He first came under the influence of Arnauld during a youthful conversion to the devout life. In 1652 his sister Jacqueline entered Port Royal and two years later, after a more profound experience of conversion, Pascal temporarily joined the solitaries. His disgust for the ways of the everyday world was compounded by the Sorbonne's treatment of Arnauld, condemned unheard by a majority more concerned with power than the theological issues at stake. As the inevitability of expulsion became clear, Pascal offered to write in Arnauld's defence. The first of 18 anonymous *Provincial Letters* appeared three weeks before the Sorbonne took its fateful decision. The series went on for just over a year, and not until 1659 was Pascal revealed as its author. Any doubt he might have had about the propriety of such worldly involvement was dispelled in March 1656 when his niece was cured of a severe eye infection by the touch of Port Royal's most prized relic, a thorn from Christ's crown. It was the first Jansenist miracle, providential enough at this moment when polemic was turning to persecution.

The first *Provincial Letter* confined itself to satirising the Sorbonne. The third, written just after Arnauld's expulsion, scornfully asserted that it had little to do with what the victim might or might not have believed. 'The sentiments of Mr. Arnauld are not heretical, but his *person*. He is a heretic not for anything he has written or said, but solely because he is Mr. Arnauld ... The grace of St. Augustine will never be true, so long as he defends the doctrine.' Grace had, in fact, been the subject of the second letter and inevitably, although they had not been directly involved in the machinations at the Sorbonne, it had brought in the Jesuits. After the expulsion, they became Pascal's main target. All the practices and doctrines denounced and deplored over 30 years by Saint-Cyran and his disciples were systematically excoriated once again, but now with public ridicule as well as scholarly venom. A new compendium of Jesuit moral theology, published by the Spaniard Escobar in 1644, provided a wealth of examples to demonstrate the laxity and (worse) modernity of their doctrines, the infinite flexibility of their casuistry, and the moral emptiness of their sacramental attritionism. 'It is truly distressing', proclaimed Letter VIII, 'to see the whole system of Christian morals overturned by such extravagances.' Characteristically, the Society did not suffer these attacks in silence and by the end of the year Pascal had provoked the intervention of the most influential Jesuit in the kingdom, François Annat (1590–1670), a notorious foe since his Latin pamphlet of 1645, *Quibbles of the Jansenists*, and now confessor to the young king. Annat accused the still-unknown author of heresy, and of speaking on behalf of Port Royal. Pascal denied both, and there were indeed influential figures in Port Royal circles who deplored this public controversy and dreaded its effects. It is certain that, by glorying in the name of Jansenist even as he poured scorn on the Jesuits, Pascal reinforced their determination to crush all those defended in the *Letters*.

If it was Pascal who wrote them, the research behind them was collaborative. Arnauld provided some material; but even more came from a vigorous newcomer who was now his constant companion. Pierre Nicole (1625–95) graduated in divinity from the Sorbonne a few weeks before the five propositions were first denounced there. Brother to a Port Royal nun, he was already teaching in the Little Schools. This remained his main occupation until 1656. After that, however, at the urging of the hard-pressed Arnauld, he devoted himself increasingly to partisan writing and research. Unlike Pascal, who gradually abandoned polemics after the eighteenth letter and died just five years later, Nicole lived until the age of 70, writing constantly on a wide range of moral and theological matters and in the end enjoying an authority the equal of Arnauld's. But he began as Pascal's ghost-writer, publishing the first collected edition of the *Letters* and a Latin translation of them for international circulation. And when they came to an end, he kept up the attack on the Jesuits by means of a further occasional series, the *Writings of the Vicars of Paris*, occasioned by the appearance in December 1657 of a Jesuit *Apologia for Casuists against the Calumnies of the Jansenists*. A work that seems less absurd today than it did to many contemporaries [36: *350–2*], these ten tracts, which were published until October 1659, were largely inspired and written by Nicole and Pascal, but they were enthusiastically endorsed by the parish priests in whose name they appeared. They argued for a vigorous and uncompromising morality, and against the 'Laxism' with which the Jesuits were wooing away their less persevering penitents [36: *135–7*]. In the end this campaign achieved a certain success. The Sorbonne condemned some of the *Apologia*'s propositions in 1658, and the next year it was placed on the index. A series of papal condemnations of Laxism would follow over the next two decades. But the price paid by the Jansenists for these triumphs was a heavy one. In the minds of Mazarin, the king, and his ministers, the link between Jansenism and the turbulent parish clergy of the recently rebellious capital confirmed all their worst suspicions. It made them all the more determined to stamp out what Louis XIV would always remember as a 'party of innovators', a 'sect inimical to all lordship, whether temporal or spiritual', which spread 'a certain spirit of independence' wherever it gained a hold [36: *349*].

[v] The formulary

Within weeks of Arnauld's expulsion from the Sorbonne, the authorities moved to close down the Little Schools at Port Royal and disperse the solitaries. They were only deterred from further action by the miracle of the Holy Thorn and the vast popular success of the *Provincial Letters*, which at their height had print-runs of 6000. So long as papal collaboration could not be relied on, such manifestations of divine and popular favour stayed the secular arm; and bishops did not feel strong enough to impose the formulary of 1655, despite its endorsement by the clerical assembly the next year. In September 1656, they wrote to Rome again, and by now the pope had decided that he must co-operate with those who governed in France. With uncharacteristic speed, within six weeks a new bull was issued, *Ad Sacram*. It affirmed that the five propositions were in Jansen's book, and condemned them in the sense Jansen had meant them. It was a quite unequivocal rejection of the distinction between right and fact.

In the spring of 1657, accordingly, the assembly of the clergy once more declared that all clerics must subscribe to the formulary. To reinforce this requirement, Mazarin insisted that *Ad Sacram* pass into French law through registration in the parlement of Paris. Predictably, the Gallican magistrates objected. They had never been asked to register *Cum Occasione* for fear of just such a reaction. Jansenism found few sympathisers within the parlement, but the magistrates were afraid that recognition of papal jurisdiction might open the way to establishing the Inquisition in France. They also feared that registering the bull would diminish their own appellate jurisdiction over the French Church. The procedure known as *appel comme d'abus* had already been used to contest attempts to impose the formulary. Only towards the end of 1657, after long negotiations had produced an explicit agreement that registration implied no extension of episcopal jurisdiction, did the parlement grudgingly accept the bull.

Persecution remained suspended during these juristic wrangles. Cautiously, the solitaries came together again to restart the Little Schools. Arnauld, against his better judgement, deliberately kept silent. But as soon as the next assembly of the clergy convened in 1660 it became obvious that a new and more determined attempt would be made to impose the formulary. The Little Schools were once more closed down, this time for good. 'I was all too good a prophet', lamented Arnauld the next year [25: *260*], 'in ever maintaining that silence was a bad remedy for the ills threatening us; and that it would have no other effect than to embolden our enemies.' The time had come, Louis XIV told an episcopal deputation at the end of 1660, to destroy Jansenism; and when, three months later, Mazarin died, he showed he meant it. The royal council ordered all bishops to impose the formulary on all their subordinates, including nuns. The same day (23 April 1661), royal officers visited both Port Royal convents and removed all their novices and boarders. The Oratorian Antoine Singlin (1607–64), chosen by Saint-Cyran himself as spiritual director of both houses, was dismissed and disappeared into hiding. These traumas no doubt played their part in hastening the death of Mère Angélique in August 1661. She at least avoided the question of whether or not to accept the formulary.

Its terms were intended to be unequivocal. It read: 'I condemn with heart and mouth the doctrine of the five propositions of Cornelius Jansenius contained in his book entitled *Augustinus*, that the Pope and the Bishops have condemned, which doctrine is not that of St Augustine, which Jansenius has explained badly against the true sense of that Holy Doctor.' Yet many Jansenists thought they could subscribe, whether unequivocally or because their primary duty was not to resist authority. There was also scope for bishops to gloss the formulary's meaning in their encyclical letters ordering their clergy to subscribe. Initially, the terms adopted by the vicars-general administering the Paris diocese (in the absence of Retz) proved acceptable to the Port Royal nuns. But they were repudiated by the crown, which demanded a signature pure and simple. Along with Arnauld, the nuns felt unable to give it without reserving their position on 'the fact of Jansen' – whether the five propositions were in the book.

A new hiatus in Franco–papal relations delayed retribution for this first direct defiance of authority by Port Royal. Retz resigned the Paris see in 1662, opening the way for resolute action by a resident archbishop of the king's choice. But, thanks to a quarrel over a diplomatic incident, no new French bishops received papal confirmation until 1664. During the interim the king sanctioned some ultimately fruitless negotiations between the leading Jansenist polemicists and certain Jesuits, but they only served to emphasise the pedantic intransigence of the Friends of the Truth, as did Nicole's renewal of public controversy with his *Letters on an Imaginary Heresy*. In 1664, the king's quarrel with the pope now over, unqualified signature of the formulary was once more required; and when none of the previous recalcitrants complied a new archbishop removed 12 of the Port Royal nuns and barred the rest from the sacraments. This produced a dozen signatures, but made the majority more obdurate than ever.

Once more the authority of Rome was invoked, and in 1665 Alexander VII was persuaded to issue yet another bull, *Regiminis Apostolici* which, for the first time, required signature on direct papal authority. But the formulary itself was subtly modified, requiring now a 'sincere' rejection of the five propositions but omitting any statement that Jansen had misinterpreted St Augustine (see appendix). In this, there was room for some compromise although, as always when recourse was had to the papal curia, there were jurisdictional objections from the parlement of Paris. Four bishops, meanwhile, issued encyclical letters allowing qualified signature, and papal threats of retribution alarmed other bishops who were otherwise perfectly orthodox.

It took the election of a new pope, Clement IX, in 1667, at a moment when Louis XIV was undertaking another war against Spain and needed to broaden his support, to defuse the situation. At the suggestion of moderate bishops trusted by all sides, the four recalcitrants were induced to apologise to a pope they had never defied, and issue new encyclical letters ordering subscription to the formulary but making no mention of right and fact. Nicole and Arnauld, meanwhile, pointed out to the non-signatory nuns that the formulary of Alexander VII did not explicitly mention the issue either. They could sign, while keeping a discreet silence as to whether the words in question were in *Augustinus* or not. They set an example by signing themselves without qualifications. Eventually the nuns did so too; and by early in 1669, this 'Peace of the Church' could be officially recognised by all parties.

Nobody was entirely satisfied. Annat, ignored by his royal penitent, sent enraged protests to a pope equally unheeding. Some nuns still agonised retrospectively; and the majority who had resisted found themselves forced to give up Port Royal in Paris to the 'jurors' who had subscribed, along with the lion's share of joint assets of their two houses. Nor were the Little Schools allowed to reopen. The sacraments, however, were once more available, and solitaries were allowed to reassemble nearby. Among them was d'Andilly, who, before he died in 1674, saw his son become foreign secretary. Even his brother Antoine, along with Nicole, was received by the king. But

Arnauld was never reinstated at the Sorbonne. And so the wounds of 25 years of controversy remained open; and the understandings of 1668 and 1669 would prove to be more of an armistice than a genuine peace.

4 Jansenism in Transition

The Peace of the Church marked the end of the classic, or heroic, phase of Jansenism. By 1669, even some of its leading protagonists were close to admitting that it was all a quibble about words. 'This generation shall pass away...' wrote Nicole in 1663, 'Other men will come, who will share none of our passions. And then it is most certain that this entire dispute will appear but a comedy, and a vain amusement... troubles so frivolous in their cause, and so pernicious in their consequences' [25: *268*]. But words, and whether or not they were in Jansen's book, were merely symptoms, or weapons, in a series of conflicts about much deeper and more substantive issues.

The theological questions remained fundamental; and although, by writing in French, Arnauld, Pascal and Nicole repeatedly sought the support of educated public opinion, the issues were doctrinal and their opponents were clerics and theologians. And what was at stake was nothing less than eternal salvation or damnation. How far were God's children free to earn their own redemption? And if there could be no redemption outside the Roman Catholic Church, how far could Catholics continue to accept an Augustinian theology now appropriated by Calvinist heretics? Calvinism was a charge repeatedly thrown at the Jansenists, and one not always easy to rebut, given the emphasis that both they and Calvinists laid upon the authority of Augustine. They also shared a moral austerity, rooted in a conviction of mankind's profound sinfulness.

Where Jansenists and Calvinists differed irreconcilably, however, was over the sacraments. In Catholic doctrine, only a consecrated priest had the power to bestow them, and all the moral authority that went with it. The Council of Trent had reaffirmed the principle in rejecting the priesthood of all believers. The priest was therefore a person of power, with real authority over his flock. He was the custodian of their conduct, which they revealed to him in the confessional. His duty was to promote their moral improvement; and accordingly spiritual direction was one of the central concerns of reforming Catholicism [13]. In this light, disputes over attrition and contrition, the frequency of communion, not to mention denial of the sacraments, were far from trivial. Nor did Jansenists have any monopoly in rigorous spiritual disciplines during this devout century [36]. The moral regeneration of the faithful was an aspiration widely shared, and along with it a deep suspicion of the flexible, smooth-tongued Jesuits who seemed prepared to bend the rules of morality to accommodate an all-too-sinful world. This was why Pascal's ridicule of casuistry and laxism struck such chords; and why many of the tenets and attitudes embraced by Jansenists commanded such widespread interest and sympathy outside their ranks.

It is possible to overestimate the solidarity and self-discipline of the Jesuits. Not all Jesuits were lax or equivocal confessors. In France there were probably even fewer than elsewhere. The more extreme and absurd examples of free-will theology and tangled casuistry treasured and repeated in Jansenist polemics tended to have been quarried from Iberian Jesuits unaware of how explosive their ideas might prove across the Pyrenees. But the Jesuits took pride in their organisation and flaunted their achievements. They thereby challenged other interests within the Church who instinctively tended to side with the Jansenists as a rallying point against the Society. The Jesuits created Jansenism, in the sense that they gave this name to a collection of doctrines and attitudes propounded by their self-proclaimed enemies. And whereas the famous definition of cardinal Giovanni Bona (1609–74) that a Jansenist was merely a Catholic who did not like Jesuits, was an oversimplification, it was based on the undeniable fact that many Jesuits tended to regard those who were not explicitly their allies as Jansenist fellow-travellers.

Nor was it only matters of doctrine and moral conduct which made these disputes more than word-play. They went to the heart of power and authority in Church and state. With their vow of obedience to the pope, their resolute pursuit of influence with kings, and their drive to become the leading educators of Catholic social and intellectual elites, the ambitions of the Jesuits seemed obvious enough. But they could only be achieved at the expense of others – whether bishops, rival orders old and new, such as the Dominicans or the Oratorians, or parish priests afraid of losing their flocks. At one time or another each of these was found siding with the defenders of Jansen's and Saint-Cyran's memory to spite the Society. Jansenists themselves, meanwhile, so well protected at Court, so well connected in legal circles, and with their own embryonic educational aspirations (or so it seemed) in the Little Schools, clearly posed a challenge to the Jesuits in more fields than recondite theology. And when, initially largely owing to pressure brought by the Society, the authority of the pope began to be deployed against opinions dear to the Port Royal circle, questions of power were transmuted into issues of jurisdiction.

If the early Jansenists had not been French, and if the main scene of their activities had not been the diocese of Paris, the consequences might have been less momentous. But, thanks to the traditions of Gallicanism, the pope could only hope to intervene effectively within the French kingdom if he had the simultaneous collaboration of the king, the bishops, and the parlements. All three had instinctive anti-papal reflexes, but because their individual traditions and priorities differed, their responses to fulminations from beyond the Alps were seldom in harmony, and it was often possible to flout papal authority by playing one Gallicanism off against another. The bishops and the parlements in particular were old jurisdictional enemies, and an *appel comme d'abus* could paralyse the exercise of episcopal authority, and beyond it, the pope's. Matters were further complicated by the virtual vacancy of the Paris see between 1654 and 1664, a decade during which the most fateful decisions were taken. Not only did this lack of a clear authority enable a convent of stubborn nuns to defy the hierarchies of both Church and state for years on end, it also allowed the parish clergy of the capital to assemble regularly for a time without supervision, noisily supporting and being supported by the leading Jansenist polemicists, denouncing the Jesuits and all their works. Even more perhaps than the *Provincial Letters*, the *Writings of the Vicars* spread news of metropolitan controversies throughout the kingdom. They were read with obvious sympathy by priests in a number of provincial cities [37: *136*]. And when at last a sustained effort was made to enforce the formulary, there were isolated instances of resistance far beyond the diocese of Paris or those of the four recalcitrant bishops [10: *113–14*]. The appeal to public opinion that had begun in 1643 with *Frequent Communion* had transformed Jansenism from a Cyranist clique, based on the Arnauld family, into a complex of religious opinions attracting widespread interest and sympathy from often divergent motives. It was not yet in any sense a movement; but the breadth of its appeal was becoming clear, and most of the ways through which it would broaden its base in later years had already made their appearance. And the Peace of the Church, so far from retarding this process, provided an interval of uneasy calm during which the roots of the 'second Jansenism' became firmly established in the soil prepared and fertilised by the first.

[i] The Peace of the Church

The Peace of the Church was underlain by ten years of Louis XIV waging or preparing for war. He had other things to think about. Besides, Port Royal, more than ever the symbolic headquarters of Jansenism, was now ostentatiously protected and patronised by the duchess de Longueville, a member of the royal family whose piety and loyalty the king respected. Such favour made the convent fashionable. The roads to it from the capital were jammed with coaches. Boarders and novices once more queued up to live in the holy precincts and join the famous community. The five propositions seemed to be forgotten; and the archbishop of Paris himself speculated on how the notorious formulary might soon be discreetly dropped.

But if respectful silence was maintained on right and fact, no such restraint was observed on other matters. Arnauld and Nicole spent the first years of the peace denouncing Calvinism, as if to show the injustice of their opponents' persistent charge that they were crypto-Protestants. They and others also devoted vast energy to publishing and popularising the works of departed leaders of their circle, such as Saint-Cyran or Pascal. The latter's *Pensées*, the unfinished fragments of a defence of the Christian religion, first appeared under Jansenist auspices in 1670. None of these works was in Latin. The controversies since 1643 had taught the Jansenists the power of the vernacular word. From there it was only a short step to the Word in the vernacular, and the 1660s and 70s saw the beginning of a major departure from Tridentine traditions with translations of the scriptures. In 1667, a team of Jansenist scholars brought out the so-called *New Testament of Mons*. It was immediately condemned by bishops, king and pope. But the next year a young Oratorian, who had just met Arnauld, produced a digest of the New Testament larded with moralising glosses. Initially called *Words of the Word Incarnate*, it was expanded and elaborated through successive editions, and became a much favoured stand-by among preachers and confessors. Even its title evolved. By 1692 it had acquired the form by which it would become famous in the eighteenth century: *The New Testament in French, with Moral reflexions on each verse to make reading more useful and meditation easier.*

The author was Pasquier Quesnel (1634–1719), and he was to become more important to the second Jansenism than Jansen himself. Like Bérulle and Saint-Cyran, he was Jesuit-educated – evidence enough, surely, that the Society's influence was not as inexorable as many feared. He joined the Oratory in 1659 and seven years later was put in charge of teaching at the congregation's Paris seminary of Saint-Magloire. It was already a centre of Jansenist sympathies, and when Quesnel arrived he found Arnauld in hiding there. The *Moral Reflexions* (as his book was to be more usually known) formed only a minor part of his writings. At Saint-Magloire he promoted the teaching of Augustinian theology and urged his colleagues to renounce their signature of the formulary. By 1676 his historical writings about the papacy were on the *Index*; and in 1684, rather than sign a new formulary, he left the Oratory and went into exile in the Spanish Netherlands, where he found Arnauld once again.

By then the Peace of the Church was at an end. Resistance to the formulary had never entirely died out in some remote dioceses and periodically royal intervention was required to silence local controversies, much to Louis XIV's irritation. Nor did he like the fact that the leading Jansenist writers continued to publish so much, and ostentatiously concert their activities; 'The king,' observed the archbishop of Paris, 'likes not what makes a stir . . . [he] wants no conventicles; a headless body is always dangerous in a state; he desires to break up all that, and hear no more of those endless people, those Port-Royal people' [12: *96*]. Orthodox bishops and the king's Jesuit confessor constantly fanned his suspicions. And when notorious Jansenists began to obstruct and denounce his other church policies, the king's exasperation was complete. In 1673, from transparently fiscal motives, he unilaterally extended his regalian rights to the whole kingdom from the limited number of dioceses where they had customarily been exercised. Only two bishops protested, but both had been among those resisting the formulary during the previous decade. One appealed to Rome, triggering a conflict between the papacy and Louis XIV that would last 15 years. Such an appeal against temporal interference in church affairs had the undisguised sympathy of the leading Jansenists in Paris. An emissary was sent to Rome to urge the pope to stand firm, and was well received by Innocent XI, flattered that those who had resisted the authority of his predecessors should now be so supportive.

It was in these circumstances, in 1679, that the duchess de Longueville died. Within weeks, the archbishop of Paris arrived at Port Royal, declaring that it was a centre of sedition. Novices and boarders were ordered to disperse at once. No more were to be accepted. Clearly the community was to be made to die out by natural wastage. It was also deprived of the priests serving it, including some distinguished solitaries. The new policy was unmistakable. The Peace of the Church was at an end; and only a month later Arnauld left France for the exile in which he was to die.

[ii] Jansenists in exile

Louis XIV now left Port Royal to wither away. He was preoccupied increasingly with extirpating a much more definite, and much more numerous, body of heretics, the Huguenots. There was also his quarrel with the pope, which culminated in 1681 in a special assembly of the clergy, which defined Gallicanism with unprecedented clarity. The four Gallican articles of 1682 proclaimed the king's absolute sovereignty within the realm in matters temporal; the superiority of councils of the Church over popes; the existence of special liberties for the Gallican church; and the need for the pope's judgement, even on matters of faith, to be acceptable to the Church. Originally drafted by the lawyers of the parlement of Paris, these articles would underpin much jurisdictional wrangling over Jansenism in the next century. Their immediate effect was to provoke an 11-year rupture with Rome, during which no French sees were canonically filled.

Although the Jansenist bishops who had initially provoked this confrontation were now dead, Jansenists continued to feel that the issues it raised were fundamental. This time they supported the pope and condemned the French bishops for cravenly sacrificing the interests of the Church to the secular demands of the monarch. Saint-Cyran would have approved. Arnauld, exulting in the freedom of exile to say and write what he liked, produced pamphlets denouncing French ecclesiastical policies. The pope even pondered making him a cardinal, and declared that Jansenism was a thing of the past.

Attitudes were certainly changing. Most of the heroic generation were now dead; and some, like Nicole, had given up the cause. Having left France two years before Arnauld, he returned in 1680 and spent his last years denouncing Protestants and other heretics. Even Arnauld's Augustinianism lost its distinctive edge, perhaps under the influence of the younger men by whom he was now surrounded. Among them were Quesnel and another former Oratorian from Saint-Magloire, Jean-Joseph Duguet (1649–1733) who, like Quesnel, had baulked at subscribing to the formulary of 1684. Duguet soon returned to France, where he lived in semi-hiding, but he kept up his links with the exiles. During these years, in fact, an extensive network of clandestine correspondence grew up among the Friends of the Truth. Port Royal, now a dwindling and isolated community of ageing nuns living on their memories, had nothing to do with it, but monasteries and chapters all over the kingdom offered discreet asylum to renegade clerics. During the years of estrangement between Louis XIV and the papacy, a semi-permanent agency of Jansenist sympathisers established itself in Rome, where they constantly lobbied the holy father. But the real nerve-centre of the network was in the Netherlands. Here, the University of Louvain was still pursuing its feud with the Jesuits and welcomed the exiles' support; while to the north, in the Dutch Republic, a large Catholic minority, not part of an established church, showed themselves more open to dissident ideas than any state church could afford to be. Quesnel often took refuge with them, as did Arnauld until he risked compromising them by denouncing William III as an usurper of the English throne from the Catholic Stuarts. The Dutch Republic was also the greatest European centre of publishing and printing, producing most of the clandestine flow of books and pamphlets whose distribution was perhaps the network's main activity.

Louis XIV's police were well aware that there was a market in France for Jansenist books and they made every effort to stem the flow at ports and frontier posts. They were not aware, however, of the extent of the Jansenist network until 1703. The shock of its discovery was to transform the king's attitude, and with it the whole character of the Jansenist problem.

[iii] Towards a Second Jansenism

The roots of this transformation lay in the 1690s. A year before the death of Arnauld, two years before that of Nicole, Louis XIV finally settled his differences with the papacy in 1693. They were never to quarrel seriously again. In 1695 the king rewarded his bishops for their loyalty during the dispute, not to mention their financial assistance in the great war he had been engaged in since 1688. He granted them complete jurisdiction over the spiritual work of their lower clergy, severely restricting the scope for *appels comme d'abus* against their authority to the parlements. The scope for jurisdictional wrangles like those of the 1650s was thus much restricted: but the parlement of Paris resented and was never to forget this diminution of its traditional authority over the Church [81: *41–2, 151–2*]. The year 1695 also saw the appointment to the archbishopric of Paris of Antoine de Noailles (1651–1729), known for his devout principles and hostility to the Jesuits. This did not make him a Jansenist. His episcopate began with a ringing condemnation of Jansenist writings. And any hint of heterodoxy would certainly have denied him the cardinal's hat which he received in 1700. But two years earlier, a pamphlet of uncertain provenance [34: *302–3*], innocuously entitled *An Ecclesiastical Problem*, had pointed out that many of the doctrines which the new archbishop had just condemned could be found in a work that he had previously recommended to all his lower clergy: Quesnel's *Moral Reflexions*.

Though often blamed on the Jesuits, this attempt to embarrass Noailles was more likely the work of Jansenists. At this moment the Society, under attack for reputedly compromising the essentials of the faith in its missionary work in China, had no interest in making new enemies. Jesuits were not involved, either, in a further attempt to revive old issues in 1702, when a new pamphlet called *A Case of Conscience* [25: *appendix iv*] argued, following a recent debate at the Sorbonne, that it was legitimate to absolve a penitent even though he maintained a respectful silence on 'the fact of Jansen' – whether or not the famous propositions were in his book. If so, the formulary of Alexander VII might be freely flouted. A new pope, Clement XI, at once condemned this position. So did Noailles. So did Fénelon, archbishop of Cambrai, anxious to re-establish his own orthodoxy after earlier flirtations with other heretical tendencies. It was at this moment, when theological controversies not seen in France for almost 40 years were flaring afresh, that Louis XIV's grandson became king of Spain and thereby ruler of the southern Netherlands. Belgian officials were authorised to arrest the leading Jansenist exiles. Among them was Quesnel; and although he soon escaped, thousands of compromising documents fell into official hands. They were shown to the Flemish Jesuits before being shipped to Paris, where more Jesuits saw them. They revealed the full extent of the Jansenist network, its links in Rome and provincial France, and the co-ordination which lay behind ostensibly random theological publications. So far from the obsession of a handful of ageing clerical eccentrics, the fight against the formulary looked more like an extensive international conspiracy reaching up to the highest levels of the Church.

The revelations galvanised Louis XIV into action. Quesnel's leading French correspondents were thrown into the Bastille. And since the renewed controversies of recent years had revealed contentious loopholes in *Regiminis Apostolici*, the best way forward seemed to be another, less ambiguous, bull. The result, after some negotiation with the pope, was *Vineam Domini* (1705), which declared that respectful silence was not an acceptable response to the questions raised in the formulary. The whole doctrinal background was also reviewed, and the main Jansenist errors once more condemned.

Although, to the fury of king and pope alike, both the parlement of Paris and the bishops raised jurisdictional objections, eventually they accepted this new bull; and Noailles was determined to have it accepted, too, by the most notorious group to have resisted the formulary, the nuns of Port Royal. In 1706, the 20 old and infirm inmates left in the rural isolation of the first Jansenism's spiritual and symbolic headquarters, were required to subscribe to *Vineam Domini*. They did so, but 'without prejudice to what was done in respect of them, at the Peace of the Church, under Pope Clement IX'. Since the whole point was to clarify the ambiguities which had made that Peace possible, reservations were unacceptable. When the nuns persisted, they were denied the sacraments. It was a sanction that had failed before, and not one to deter those who believed themselves seldom worthy to take the sacraments in any case. The nuns disavowed in advance any acceptances that might be wrung from them by force – and none succumbed. Accordingly in 1709 they were all dispersed to other monastic houses, and the original Port Royal was formally closed down. Two years later, to prevent the site becoming a centre of pilgrimage, the buildings were razed and the remains of 3000 people disinterred. Many had lain there since long before Angélique Arnauld had seized control of the convent a century earlier; but the bodies of hundreds of Friends of the Truth, including most of the great names of the heroic age, were now dug up, most to be reburied in a common pit.

Even the pope, and some of Jansenism's most resolute critics such as Fénelon, had not wished to see such a posthumous mass martyrdom. The responsibility lay clearly with Louis XIV himself, and perhaps with his last and most inflexible Jesuit confessor, appointed in 1709, Michel Le Tellier (1643–1719). Le Tellier had spent much of his career denouncing and working against Jansenism; and he was not content with the mere physical destruction of its symbolic centre. He wanted a final and unambiguous condemnation of every doctrine associated with this nest of heretics. Nor was he alone. His fellow Jesuits had shown the way when, in 1704 and 1705, they had continued the quarrel over the *Case of Conscience* in pamphlets declaring Quesnel a seditious heretic, and his *Moral Reflexions* nothing less than a compendium of all the Jansenist errors. In these terms they were denounced to the curia in 1707, and the book was formally condemned by the Inquisition. This, of course, meant nothing in France; but in 1710 Fénelon induced two bishops to issue pastoral condemnations of the *Moral Reflexions*, and to publish them within Noailles's diocese. This book, they said, was 'full of impious dogma, where all the errors and maxims of the Jansenists are taught on every page' [34 : *318*]. Everybody now knew that it was a work that the cardinal archbishop of Paris had once commended, and Noailles was outraged at this blatant attempt to embarrass him. But the protests he made to the king had exactly the effects that Fénelon had hoped. Louis XIV, exasperated by the persistence of a problem he had expected to solve at the start of his reign, was persuaded by Le Tellier that this was the opportunity for a final and doctrinally authoritative condemnation of all that Jansenism stood for in the works of its apparent leader, now an exile among Dutch heretics in an enemy state. Ignoring an unpropitious string of precedents, he asked the pope for one last bull.

5 *Unigenitus* 1713–32

Clement XI was very reluctant to issue yet another bull, but he could scarcely refuse a direct request from Louis XIV. All he could do was delay, while a congregation painstakingly combed Quesnel's book for compromising statements. In the end it found 101 of them, which were formally condemned by the bull *Unigenitus*, issued on 8 September 1713 [text in 6, ii: *305–41*].

These propositions, the pope declared, were 'false, fallacious, offensive, injurious to pious ears, scandalous, pernicious, rash, damaging to the Church and her customs, outrageous not just to her, but to secular powers; seditious, impious, blasphemous, under suspicion of heresy, reeking of heresy, favourable to heretics, heresies and schism; erroneous, close to heresy, and often condemned; in fine heretical, renewing divers heresies, principally those contained in the famous propositions of Jansenius, taken in the sense in which they have been condemned'. All the faithful were forbidden to think about, teach, or speak of these propositions, except to refute them; and they were forbidden to read, copy or use the *Moral Reflexions* on pain of excommunication.

Forty-three of the propositions concerned grace and predestination, rehearsing in various ways the substance of the original five of 1649. Another 31 (44–71, 87–9) dealt with the operation of divine love or charity, and the deplorable results of human self-love or concupiscence. These concepts were central to the thought of St Augustine, and instantly recognisable to the trained eye. Some underlaid the practice of contrition. Most of the propositions condemned, therefore, evoked the quarrels of the 1640s and 50s. But others were clearly related to what had happened since. Seven (79–85) recommended the reading of the scriptures and more active participation in Church affairs by the laity. Seven more (72–8) implied that the Church existed independently of the hierarchy. Most of the rest criticised in one form or another the measures taken by authority to stifle dissent by excommunication, the imposition of oaths, and other persecutions. They implied that these actions had been unjust, unjustifiable, and beyond the jurisdiction of those taking them.

No pope could accept such imputations once they were drawn formally to his notice; and Louis XIV, in requesting the pontiff to condemn them in a bull, had assured him that this time there would be no problems. It proved a massive miscalculation. Not only did *Unigenitus* infringe the prized liberties of the Gallican church on all the traditional grounds (now reinforced by the king's own four articles of 1682); it also made new claims. Matters hitherto entirely within secular competence, such as what French subjects might or might not read, were now claimed for the Church. The bull also appeared, in condemning Quesnel's critique of excommunication, to be asserting the pope's right to override a subject's duty to the king. Louis XIV clearly believed that his own endorsement of a measure he had himself requested was enough to ensure its passage into law. His leading jurists demurred. They thought it beyond his power to surrender the traditions of the French Church and his own authority. 'No constitution of the popes in matters of doctrine', declared his procurator-general at the parlement of Paris, Henri-François d'Aguesseau (1668–1751), 'may be invested with the king's authority without a legitimate and sufficient acceptance by the Gallican church; to do otherwise would be to recognise the pope as infallible and sole judge of faith against the fundamental maxims of our liberties.'

The public uproar was unprecedented since the king's minority. In 1714 alone, around 200 books and pamphlets were published on the topic. Debates on *Unigenitus* reduced the Sorbonne to chaos. The parlement of Paris, despite spectacular threats from the enraged king, refused to register the bull until it was accepted by the bishops. Eventually it yielded, but its registration was qualified by a rider safeguarding Gallican liberties. And when, as under Mazarin in 1655 (see above, p. 27) a less-than-plenary meeting of the episcopate accepted the bull, the magistrates declared it inadequate. In any case, the bishops were divided. A minority, led by Noailles, demanded further clarifications. Meanwhile, the cardinal forbade publication of the bull within his diocese. By the summer of 1715 the king had decided to convene an unprecedented full council of the French Church to accept the bull once and for all and depose Noailles. He was preparing to force the parlement to register this summons when he died.

[i] The mainsprings of opposition

The reaction against *Unigenitus* showed that Jansenism now enjoyed extensive support among the clergy of France. Among the bull's opponents, a correspondent told the pope, were 'All the fathers of the Oratory, the Benedictines, nearly all the monks of Cîteaux and of St Bernard, canons regular, the Dominicans, and a great number of other monks and religious of other orders, together with almost all the secular priests' [6: *250*]. Certainly, by 1718, around 7000 clerics were prepared to support the manœuvres of Jansenist bishops [49: *202*]. The number of lay supporters is impossible to estimate, but must have been even higher if the impressions of contemporary observers are any guide. What chords had *Unigenitus* struck, to turn a dissident fringe into such a widespread movement of protest?

Because Jansenism represented an increasingly diverse range of tendencies and attitudes, the very attempt to produce a comprehensive condemnation inevitably reinforced what welded them together. Focusing on Quesnel's book compounded the effect. Here was a devotional work, with the New Testament as its core, widely issued over 40 years, and often commended by episcopal authority. That it had been compiled by a persecuted exile who had devoted his life to the memory of men like Arnauld and Saint-Cyran was beside the point. The *Moral Reflexions* preached an austere, devout morality and the primacy of spiritual values in everyday life. It was a serious call to Christian self-improvement which confessors with no interest in the 'fact of Jansenius' found well suited to the needs of their penitents and parishioners.

Or rather, it suited what a new generation of priests conceived those needs to be. For, if the most satisfactory general definition of Jansenism is resistance to the orientation given to Catholicism by the Council of Trent, one Tridentine principle proved fundamental to its spread. The professionalisation of priestly training through the establishment of seminaries was a slower process than anyone foresaw; but by the later seventeenth century, and especially in France, it had at last gathered momentum. Between 1642 and 1698, 104 seminaries were established throughout the kingdom. As a result, the parish clergy confronted by *Unigenitus* were very different from those subjected to the formulary two generations earlier. Most of them had now received a formal clerical education which had stressed two things above all. One was the pre-eminent duty of pastoral care, where Quesnel's book, as Noailles put it in his celebrated recommendation, was 'worth a whole library'. And this training had led them to believe that the faithful would benefit from reading the scriptures at every opportunity. Secondly, their training had imbued them with a sense of the special destiny of ordinary priests – that same sense which inspired the Oratory from which Quesnel came. Holy orders, and the obligations they imposed, took precedence over the temporal authority of the hierarchy, whether of bishops (despite the enhanced powers granted them in 1695) or the holy father himself. Such convictions made the ordinary clergy very receptive, like their *frondeur* predecessors in Paris in the 1650s, to the ideology of Richerism and clerical democracy, which seemed to be the target in the condemnation of propositions 72–8 and 90–1. None of this meant that a majority of parish clergy ever opposed the bull; but it does help to explain the motivations of those who did.

Their Jansenism was moral, pastoral and professional rather than theological. Doctrine had more to do with the opposition to the bull found among regulars. Some, like the Oratorians, shared the parish clergy's pastoral concerns, but the contemplative life to which most were vowed allowed more time for intellectualising. Orders committed to scholarship, such as the Benedictines, were inclined, like the early Jansenists, to regard the legacy of scholasticism as corrupt, and to seek the often subversive purity of the original texts. The Dominicans, committed as ever to doctrinal rectitude, noticed with approval that Quesnel, Duguet and other third-generation Jansenist writers had moved in detail closer to their beloved St Thomas Aquinas than to Calvin, or indeed Jansen. And so, many regulars, whose co-operation had been crucial in the condemnation of Arnauld in the 1650s, were more inclined 60 years later to support his self-proclaimed heirs. Monastic asylum had been instrumental in the survival of Jansenists when they were driven underground after 1679. Enclosed, self-recruiting, and organised into kingdom-wide networks, once the virus of Jansenist sympathies was implanted in any of these orders it spread very easily, and was extremely hard to eradicate [10: *113–18*]. Though more open, the Oratorians maintained their unity through their own seminaries, such as the notorious Saint-Magloire, where Arnauld had hidden, Quesnel had taught, and despite acceptance of the formulary in 1684, Augustinian doctrines were resolutely transmitted.

And one thing continued to unite seculars and regulars as it had from the start: dislike of the Jesuits. During Louis XIV's quarrel with the pope, the influence of the Society was somewhat eclipsed. By the time the rift was healed, a great controversy over their willingness to allow special rites to Chinese converts was besmirching their reputation once again. In one of the last things he wrote, Arnauld alleged that the compromises made by the Jesuits with Chinese habits were typical of their lack of doctrinal scruple. The Sorbonne condemned them; the Dominicans sought to undermine the Society in Rome. But Clement XI temporised on this issue, and the promulgation of *Vineam Domini* was seen as a sign of Jesuit recovery in both papal and royal esteem. Everything that happened between 1705 and 1713, including the brutal and vindictive treatment of Port Royal and its nuns, looked like an intensifying Jesuit revenge on tormentors too long unpunished. *Unigenitus* brought it to a climax; but it also brought together all the alarmed elements of the traditional, instinctive anti-Jesuit coalition in France.

No doubt they derived comfort from the knowledge that Louis XIV's long reign must soon give way to a regency. Jansenism had thrived under the weak and fluctuating rule of the previous regent. Now it was to do so again. The intellectual foundations were, however, established during the initial protest against the bull, or 'constitution' as it was usually called. At that point, the *Hexaples, or six columns on the Constitution Unigenitus*, in which texts from scripture and the fathers were juxtaposed with the condemned propositions to prove their orthodoxy, was first produced. It was to be constantly added to over subsequent decades. Another key text was Vivien de La Borde's *Witness to the Truth* (1714), which castigated the hierarchy for betraying the truth, upheld the rectitude of a faithful minority as the true voice of the Church, and vaunted the validating witness of the laity – that public opinion to which Jansenists had always appealed since *Frequent Communion* [81: *77–9, 93–5*]. La Borde found hidden meanings in scripture, a mode of interpretation known as figurism. Taught by Duguet, and his disciple at Saint-Magloire, Etemare (1682–1770), figurism was a creed for a persecuted minority [63; 49: *200–2*]. The Bible had foretold, figurists argued, a time of mass apostasy before the final deliverance. But God had not 'cast away his people, which he foreknew ... a remnant according to grace' [Rom. xi: 2, 5]. Their duty was to safeguard the truth during trials which would precede its ultimate triumph.

Figurist writers have now been shown to be the animators of the first outburst of publicity against *Unigenitus*, and of much of what followed later. Numbering no more than 60, and concentrated over-whelmingly in Paris, their messianic zeal was subsidised by a special fund to which all Friends of the Truth were encouraged to make donations and bequests. Founded by Nicole and named after the servant to whom he entrusted it, 'Perrette's Box' was worth several hundred thousand livres by the mid-eighteenth century, and continued to grow even after Jansenist numbers began to dwindle [63: *131–5*].

[ii] The Appeal

The Regent d'Orléans, who took power in 1715 on behalf of the five-year-old Louis XV, was religiously indifferent. Plans for a national council were abandoned and Le Tellier ordered to retire. Jansenist prisoners were released, exiles allowed to return. Religious affairs were entrusted to a 'council of conscience' presided over by Noailles and packed with notorious Gallicans. Many who had accepted *Unigenitus* under the old king now recanted, including the Sorbonne and a number of bishops. Other prelates urged the regent to seek further clarification from Rome. But Orléans's neutrality cut both ways. Orthodox publicists denounced Jansenism in a ferocious series of pamphlets called the *Tocsins* (calls to alarm). A majority of the bishops stood by their acceptance, and some tried to force their lower clergy to conform with threats of excommunication. The parlement of Paris ominously forbade bishops to extract statements of acceptance until the bull had been accepted in a proper Gallican way – an implicit disavowal of the earlier registration. Yet the pope remained obdurate. He refused to explain or clarify anything further, and withheld confirmation from new bishops suspected of opposing his bull. It was a judgement of faith, and thus irrevocable.

Polemical publications continued to pour from the presses: they had reached a thousand titles by 1730. In 1716, Duguet produced a figurist handbook in the form of *Rules for the Understanding of the Holy Scriptures*. The jurisdictional aspects of the conflict were reviewed the same year by Nicolas Le Gros (1675–1751) in his *Of the Overthrow of the Liberties of the Gallican Church in the affair of the Constitution Unigenitus*; but he went beyond legalities to vaunt, in Richerist language, the independent authority of ordinary priests as judges of faith and peers of the bishops in church government

[81: *79–81*]. And the argument was raised to a new level of principle in March 1717, when four bishops who thought the constitution beyond satisfactory explanation played the ultimate card in any conflict with the papacy. In a solemn appearance before the Sorbonne, they appealed against *Unigenitus* to a future general council of the Church.

No such council was to meet until 1869, and by then nobody was concerned about *Unigenitus*. The problem for the 'Appellants' was that only the pope could convene a council. In that sense their gesture was hopeless. Nevertheless the excitement carried the 'anti-constitutionaries' along. The Sorbonne endorsed the appeal by a huge majority. Noailles subscribed to it, although secretly at first. So did a number of other bishops, several influential chapters and two thirds of the diocesan parish priests of Paris [65]. The effect was to destroy three years of attempts by the regent to stand above the battle. It was clearly impossible for the crown to remain neutral, with Rome insisting that it honour Louis XIV's promises, but the number of the bull's opponents seemingly growing. A first attempt to impose a solution only made matters worse: while the four bishops were exiled to their dioceses, further publications on either side of the controversy were forbidden. Glimpsing resolute action at last, the pope intervened. In 1718 the Inquisition condemned the appeal and he excommunicated the four bishops. The encyclical *Pastoralis Officii* (1718) renewed the instruction to all the faithful to accept *Unigenitus* on pain of excommunication. But any new papal instrument raised all the old jurisdiction problems with the parlement, already at loggerheads with the crown on a range of other issues. Controversy exploded again as Noailles made his appeal public, and (emancipated from any conciliatory role by the abolition of the council of conscience) launched a new appeal against the latest pretensions. Ten thousand copies of his declaration were sold and hundreds of clergy who had hesitated to subscribe to the original appeal of 1717 now did so. It was the numerical high-point of Jansenism, when perhaps 10 per cent of the French clergy rallied publicly against *Unigenitus*, and now no less than three-quarters of the parish clergy of the Parisian archdiocese.

Once more the regent sought to recover control. In June 1719 he decreed yet another silence on all religious matters for a year. This truce produced a fudged form of words, the so-called *Body of Doctrine*, which the ever-irresolute Noailles accepted, along with most bishops. Even the carping parlement, now in exile at Pontoise and anxious to return, seemed open to negotiation. By the end of the year it had grudgingly agreed to register a royal declaration incorporating the *Body of Doctrine* – although with the same Gallican reservations as in 1714. By then, however, a further determined attempt had been made to torpedo any new peace of the church by the four original appellants. In November 1720 they defiantly renewed their appeal to a future council.

And so the schism continued, with parish clergy flocking once more to subscribe the new appeal. Tellingly, however, they were not as numerous as before; nor was the parlement deterred from registering the new declaration by this repeat of an old manœuvre. The movement against *Unigenitus* had in fact peaked. Leading figures in the quarrel were now dying off: Quesnel in 1719, Clement XI in 1721. A further vacancy in St Peter's chair only three years later meant that no clear policy emerged from Rome until the late 1720s. In Paris, meanwhile, ecclesiastical policy was put into the hands of a committee of pragmatic but orthodox prelates inclined to action rather than declaration. Their first step was to go back to a tried baseline. In 1722 all clergy were required to subscribe, for the first time in many years, to the formulary of Alexander VII. Prominent in this approach was the king's former tutor who, born in the year of *Cum Occasione*, could remember the Peace of the Church, during which his own ecclesiastical career had begun. Later, as a bishop, André Hercule de Fleury had made his position clear to no less a person than Noailles: 'I am quite convinced that the decisions of the Church alone should guide us in everything... those who are truly obedient [should] come together and do no more than defend her decisions without entering into contested decisions. That is the only way to support her, as we ought, for there will be no end to it otherwise' [49: *43*]. These principles would guide him when, between 1726 and 1743, he was the effective ruler of the kingdom.

[iii] Persecution, publicity and lay intervention

So long as Noailles remained archbishop of Paris some protection for Jansenists was assured. Chronically irresolute, he could never be induced to deploy his diocesan authority against appellants and their sympathisers. The character, or lack of it, of the archbishops of Paris was crucial to the entire history of Jansenism. None of Noailles's fellow metropolitans, however, shared his scruples; and when in 1726 one of the four original appellants, bishop Jean Soanen (1647–1740) of the tiny Alpine diocese of Senez, issued a pastoral letter denouncing *Unigenitus* and recommending the faithful to read Quesnel, the archbishop of Embrun haled him before a provincial council in 1727, which suspended him and exiled him to a remote abbey. There he remained until his death 13 years later, a living martyr in the Saint-Cyran mould.

Soanen was that rare thing, a non-noble bishop without connections at Court, and habitually resident among his remote flock. For all these reasons he seemed an easy example to make. But he was also noted for his piety and pastoral commitment, whereas his archbishop was an absentee and a notorious rake. And coming as it did on top of a series of other suspensions and exiles among lesser recalcitrant clergy, Soanen's fate awakened public awareness that a more determined policy was now in operation. Nothing did more to concentrate this awareness than the publication in the capital of a legal brief in his favour, denying the legality of the Council of Embrun and its action against him. It was signed by 50 advocates at the parlement, and orchestrated by a smaller number with long experience of advising on legal resistance to *Unigenitus* [48: *72–88*; 49: *206–13*]. At a time when clerical opposition to the bull was dwindling under the impact of Fleury's persecution (an attempt to organise a second re-appeal in 1726 collapsed from lack of support) this *Consultation of the Fifty* marked the start of a new phase in which leadership of the Jansenist cause would begin to be taken over by laymen, and its success depend on lay support as never before since the days of Pascal.

To maintain lay support it was essential that the public never lose sight of the issues. Occasional illicit pamphlets and printed briefs (not subject to censorship) could perform that function when contested cases arose, but Jansenist zealots perceived that some more regular stimulus was needed. They found it in a newspaper, the *Nouvelles Ecclésiastiques* (Ecclesiastical News), which first appeared on 28 February 1728. Initially modelled on the manuscript handbills that had long circulated among the Friends of the Truth to keep them aware of the multifarious threats it faced, the *Nouvelles* rapidly established itself as a journal of record and polemic read far beyond the ranks of Jansenism. Printed, published and distributed secretly, it was to continue appearing right down to 1803, defying every attempt by successive governments to track down its authors and printers. Its very success in eluding the best police system in Europe made it legendary and won it a readership of intense loyalty. And from the start its authors were determined to demonstrate, and thereby foster, the widest possible lay support for the appellants by recounting every incident of popular Jansenist piety [52: *63–81*]. There was certainly enough to report in 1728. One observer recorded that most of Paris was Jansenist, with only bishops, clerical careerists and the Jesuits and their pupils prepared to accept the bull. Women, he repeatedly noted with astonishment, were particularly vehement against it. Even his own manservant asked him if confession would be abolished if the bull were accepted.

So there was general amazement when Noailles, now close to death but pressed relentlessly by Fleury, issued a pastoral letter which appeared to accept *Unigenitus* unequivocally. Then, characteristically, he disavowed it. Fleury was visibly relieved when Jansenism's most persistent, if inconstant, protector finally died in May 1729. On the same day he nominated a successor of proven orthodoxy, archbishop Vintimille of Aix (1655–1746). Clearly Fleury thought the last obstacle in his campaign to purge the clergy of Jansenism had disappeared. Many wavering fellow-travellers, such as the canons of Notre Dame, obviously thought so too and hurriedly voted to accept a bull they had long resisted. Nor was it long before the new archbishop began to demand that others follow their example. A recalcitrant Sorbonne was simply purged of opponents by royal order. Parish clergy with equivocal records were suspended; tainted colleges and seminaries, including Saint-Magloire, closed down or transferred into safer hands. Shrill protests from appellants hitherto protected by Noailles were ignored; and early in 1730 Fleury decided to consolidate these advances for orthodoxy by resuming the policy left unfinished by Louis XIV. A declaration of 24 March proclaimed *Unigenitus* a 'dogmatic judgement of the Universal Church by virtue of the general consent of the episcopate', and therefore a law of the state, too. All clergy were required to accept it. Appeals to the parlements against punitive action by bishops were expressly forbidden, reinforcing the extension of episcopal authority granted in 1695 (see above, p. 42).

Fleury was well aware, of course, that laws of state must be registered by the parlements. Extensive private consultations with leading magistrates had therefore preceded the promulgation, and one result was the description of *Unigenitus* as a dogmatic judgement of the Church rather than (more provocatively) a rule of faith. Even these precautions did not prevent a handful of Jansenist magistrates, led by the clerical counsellor Pucelle (1656–1745), from trying to obstruct registration. It took a *lit de justice* in the king's presence to override them. The triumph of orthodoxy seemed complete, however; and bishops, with more confidence than ever before, moved to enforce it on the rapidly shrinking remnants of the appellants. They were assured that any appeal against their authority would be evoked to the king's council and never reach the parlement.

But defeating Jansenism was no longer simply a question of bullying parish priests and unworldly regulars. The laity was now involved in ways not previously seen. The advocates who had produced the *Consultation of the Fifty*, for instance, rushed to write briefs for priests still determined to launch appeals. Their legal arguments, rather than doctrinal ones, were to dominate quarrels over Jansenism for the next generation [47: *248–58*]. A new consultation issued late in 1730 challenged the very foundations of political authority by asserting that laws were based on a contract between governors and governed [46: *3–6, 91–104*]. Attempts by shocked ministers to suppress it outraged both the corps of advocates and the parlement, and in August 1731 a full-scale strike of the bar was declared after an avalanche of fruitless remonstrances from the parlement. It scarcely seemed a coincidence when, that same month, God Himself appeared to intervene on the side of his elect to mobilise the support of ordinary believers.

[iv] Miracles and convulsions

Although Jansenism was rooted in a sense of God's infinite remoteness from a world polluted by Adam's sin, Jansenists at bay had always watched desperately for evidence of divine intervention amid their tribulations. Thus, the miracle of the Holy Thorn (see above, p. 29) had served to stay Mazarin's hand in 1656, and there had been other supernatural signs at Port Royal during its first resistance to the formulary. The figurist tone of much Jansenist writing after the destruction of the convent aroused expectations of further signs from God; and throughout the late 1720s there were reports of miraculous cures experienced by pious invalids in parishes served by Jansenist clergy. They helped to sustain the faith of Noailles himself during his wavering decline, and his public endorsement of some lent them official credibility. Among marvels boosted in this way were cures said to have occurred at the tomb of François de Pâris, an ascetic graduate of Saint-Magloire, renowned for his piety, charity and opposition to *Unigenitus*. He had died professing the Truth in 1727 and was buried at Saint-Médard, a poverty-stricken parish in the east of the capital. Trumpeted by the *Nouvelles Ecclésiastiques*, his tomb soon became a place of pilgrimage. And although Fleury prevented a full confirmation of the miracles by the dying Noailles, the cult continued to flourish.

The number of alleged cures began to grow during the process of making *Unigenitus* a law of the state. One spectator who came to scoff at the devout scenes around the tomb was stricken with paralysis. Miraculous recoveries, meanwhile, were increasingly preceded by spectacular writhings, convulsions and speaking in tongues. Soon those hoping for a cure were positively expected to convulse, and the cemetery was turned into a quagmire by the feet of thousands who came to watch. Richer visitors could hire ringside seats at what more fastidious observers were soon comparing to a circus. The convulsions at Saint-Médard, in fact, were to prove a milestone in the development of the French Enlightenment in demonstrating the ignorant credulity and absurd fanaticism that religious passion could unleash.

Even some sincere Jansenists thought scenes that would surely have disgusted Saint-Cyran and Arnauld (though less certainly Pascal) an embarrassment, and damaging to their cause. As time went by this would become their predominant view. Yet in 1731 many more saw evidence of divine intervention against the bull. A magistrate from the parlement, L. B. Carré de Montgéron (1686–1754) was converted from a rakish life by what he saw and devoted the next six years to writing a pious chronicle of the miracles. Even he could find no more than eight indisputable ones among the mass of hysterical fits and holy-rollings, but it was more than enough to determine him to lay the record before the king in person. When, in 1737, he eventually inveigled himself into the royal presence and presented his book as an account of truth kept hidden by wicked ministers, it cost him his freedom. More practical and experienced Jansenists used the miracles as a further opportunity to test the archbishop's authority. The *Nouvelles Ecclésiastiques* demanded that he begin procedures of authentication. When he refused, and condemned the cult, the number of reported cures shot up [61: *78–80*]. And the refusal, not being directly connected with *Unigenitus*, gave grounds for appeal outside the edict of March 1730. Repeated evocation of previous appeals to the royal council had already exasperated the parlement at a moment when the bar was also on strike over related issues. In the autumn of 1731, quarrels arising from *Unigenitus* seemed to be reducing the public life of the capital, and with it the mainsprings of political authority in the entire kingdom, to chaos.

Ever since his precarious success in getting the bull registered as a law of state, the guiding principle of Fleury's policy had been to do nothing that would further inflame a situation already delicate and complex enough. Vintimille complained constantly that he did not get the ministerial support he deserved. But now, Fleury recognised that he must act. First, he ended the 'affair of the advocates' by an ambiguous form of words which allowed them to claim a victory [48: *102–5*]. Then, as they basked in success, he moved rapidly to close the Saint-Médard cemetery, not on grounds of religion, and not by the archbishop's always-challengeable authority, but as a police measure, to preserve public order [59: *ch. 5*]. This took place on 29 January 1732, and there was no resistance. A huge show of force ensured that, after several days of milling outside the locked gates amid much candle-bearing and psalm-singing, the crowds of thwarted pilgrims melted away.

> God take note, by royal command:
> Miracles in this place banned

said an irreverent placard found outside. It was indeed the end of Jansenist miracles. Convulsionism continued, but now in private, developing strange sado-masochistic rituals which sufferers claimed brought them relief. Convulsionist cells lingered on until the 1780s and in the provinces even later [61, 63]. But few Jansenists any longer found evidence of the hand of God in what they did. The reinforcement which the Saint-Médard episode brought to the campaign against *Unigenitus* was in fact fleeting. Now deprived of a spectacular focus of attention, popular interest in the arcane concerns of theologians and jurists evaporated. At the same time, the declining strength of opposition to the bull, already evident in the 1720s, stood out once more, now compounded by divisions about Saint-Médard itself. As the generation of seminarians who had responded to the appeals began to die off, only serious miscalculations by the authorities were likely to arrest that decline.

6   Refusal of Sacraments 1732–60

Between the closure of the Saint-Médard cemetery and his death 11 years later, Fleury almost destroyed the second Jansenism. His policy was simple: to use the crown's extensive powers of ecclesiastical patronage to ensure that, whenever a benefice fell vacant, it was filled by a priest who accepted the 'constitution'. Fleury kept the *feuille des bénéfices*, the office of ecclesiastical appointments, firmly in his own hands. Plagued throughout his ministry by the Jansenism of long-serving episcopal renegades like Soanen, or his fellow-appellant the well-connected (and therefore untouchable) bishop Colbert of Montpellier (1667–1738), the cardinal was determined only to create bishops of proven orthodoxy. But orthodoxy alone was not enough. It was equally important, especially in the case of bishops with extensive jurisdictional powers enhanced by the edict of 1695, to appoint men of sound judgement and moderate inclinations, who would not provoke needless confrontations.

What could happen otherwise was shown in 1732. Exulting in the ease with which the Saint-Médard cemetery had been closed, in May Vintimille issued a pastoral letter condemning the *Nouvelles Ecclésiastiques* and instructed his diocesan clergy to read it from the pulpit. Twenty-one, all former appellants, refused. They appealed to the parlement, but like so many other appeals since 1730, it was evoked to the royal council.

Their exasperation at repeated denials of jurisdiction fanned by the arguments of a handful of Jansenist colleagues [49: *249–50, 321–5*], the magistrates felt it was time to take a stand. 'The parlement is convinced', wrote a well-informed observer, 'that the king wanted to deprive it of all ecclesiastical affairs, of all appeals ... and make the bishops into absolute masters of their dioceses' [49: *261*]. A summer of confrontations ensued, with magistrates arrested, judicial strikes, remonstrances and mass resignations. The government responded with a sweeping curtailment of the parlement's jurisdiction and right of remonstrance. The original issues were virtually lost sight of in what became a struggle of authority between the king and his leading law court. In the end both became alarmed by the extremes to which they were pushing each other; and in December complex negotiations produced a resumption of functions by the parlement in return for an agreement by the king to suspend the new restrictions on its powers [49: *chs 11 and 12*].

The Jansenists who had provoked this crisis were happy to keep their cause in the public eye when there were no more miracles to capture popular attention. Most magistrates were much less pleased to be manipulated by a determined minority, no more than 14 strong but extremely vocal. But in any case, from 1733 onwards a new war diverted attention from domestic issues. Throughout the rest of the decade periodic attempts were made to engineer further crises, but the parlement was reluctant to become so dangerously embroiled again. Even when Carré de Montgéron, one of its own members, was arrested in 1737, its protests were half-hearted. And meanwhile Fleury proceeded relentlessly with his purge of the clergy, steadily packing every available benefice in his gift. Unfortunately, not every prelate known to be sound on *Unigenitus* could be relied on to enforce orthodoxy circumspectly. Nor could all those whom they in turn presented to, or confirmed in, lower benefices. Over-zealous parish clergy, in fact, were soon proving even more provocative than their bishops in the way they chose to persecute their opponents through the sacraments.

[i] Refusal of sacraments

Denial of the sacraments already had a long history in the struggle against Jansenism. Although it was a basic Jansenist spiritual strategy not to profane the sacraments by seeking them too often with insufficient preparation, this was the corollary of a veneration which made their total loss unbearable. Nothing had distressed the nuns of Port Royal more than their virtual excommunication during long bouts of persecution. The uncertainties of the regency, and the mass support given to Jansenists during the 1720s, made any attempt to enforce *Unigenitus* by refusal of sacraments very hazardous. But once the constitution was registered as a law of the state, clerical zealots took the offensive and throughout the 1730s suspected opponents were repeatedly denied communion. Several congregations of recalcitrant nuns found themselves repeating the experience of Port Royal at the hands of intransigent spiritual directors. The most notorious victim, perhaps, was Carré de Montgéron, boycotted by local priests in his exile until Fleury discreetly intervened. But not all sacramental damage could be so easily undone. Denial of extreme unction, the last sacrament, meant that a sinner died unabsolved and would be denied Christian burial. Yet the very severity of this sanction lent it unmatched power, and within months of the registration a dying woman in Orléans was refused the last rites upon denouncing *Unigenitus*. The evocation of an appeal to the parlement against this alleged abuse of priestly authority was one of the major issues leading to the judicial crisis of 1731–2. In its aftermath, Fleury was anxious to prevent further provocations, and this policy went hand-in-hand with his purge of the clergy. By the early 1740s, beneficed clergy with a Jansenist record were a rarity, and those left were dying off. Fleury was content to let them do so in peace. But by the time he died himself in 1743, it was becoming clear that if anything his earlier efforts had succeeded too well. A younger generation of priests, trained in seminaries scoured of questionable doctrines, was outraged to see notorious flouters of the Church's authority still entrusted with the cure of souls. They were no less concerned by the freelance competition of unbeneficed and unauthorised opponents of the bull for the confidence of the laity. Fleury's careful and cautious approach seemed to them too indulgent. But, in seeking to finish off Jansenism by methods swifter than steady erosion, they goaded it into a further spasm of explosive life.

[ii] Confession notes

When Fleury died, Louis XV assumed personal direction of his government. He entrusted ecclesiastical patronage to his son's preceptor, J. F. Boyer (1675–1755), bishop of Mirepoix, an unworldly former monk without Fleury's awareness of the wider implications of his appointments. The most crucial of these was made when the see of Paris fell vacant in 1749. It went to Christophe de Beaumont (1703–81), archbishop of Vienne, who had still been a child when *Unigenitus* was promulgated, and had trained for the priesthood during the dangerously schismatical days of the appeal. His commitment to the welfare of his flock was beyond doubt. He was a devout man of conscience. But if 'devoutness' (*dévotisme*) had been one of the sources from which Jansenism sprang in the seventeenth century (see above, p. 15), by the eighteenth it meant militant defence of the Church's authority against all comers, including Jansenists. Beaumont was to become the figurehead to the new devout party, by trying to complete what Fleury and Vintimille had begun.

A new intransigence was already in the air before his appointment. In the diocese of Amiens, clergy who suspected the orthodoxy of dying penitents were authorised to refuse them the last rites unless they produced a certificate (*billet de confession*) signed by the priest to whom they had last confessed. The device was not entirely unprecedented. Its object was to flush out unbeneficed and unauthorised confessors. Hardly had Beaumont assumed his see in Paris when his lower clergy began to introduce the same procedures against Jansenist suspects. The first victim was Charles Coffin (1676–1749), appellant, college principal and former rector of the university, denied the last rites because on his deathbed he would not produce a confession note. The archbishop backed the denial of sacraments and Coffin was left to die unshriven.

Four thousand people attended his funeral. Many were former pupils, but this was more than a demonstration of personal loyalty. It was a protest against clerical despotism. To demand acceptance of a disputed body of doctrine in the last moments of life, through the fear of dying in sin, seemed an outrageous abuse of priestly authority. It came at a moment, too, when the bishops were orchestrating resistance to royal plans for subjecting the wealth of the clergy for the first time to direct taxation [64: *83–7*], so that the Church appeared to be in the hands of greedy and selfish power-seekers, indifferent to the desires or interests of the faithful laity. The roots of later Parisian anti-clericalism, which would become virulent in the Revolution, can perhaps be found in the series of episodes that now followed from refusal of sacraments.

What made them serious at the time was the renewed involvement of the parlement. After half a dozen cases had occurred, it complained formally to the king. But by the time of these remonstrances, in 1751, a further complication had arisen. The archbishop was determined to eliminate Jansenists from the administration of the federation of Parisian poor-houses and asylums, the General Hospital (*Hôpital Général*). Attempts to place the hospital in the hands of his own nominees were resisted by Jansenist lawyers on its governing body, who eventually carried the appeal to the parlement [69: *83–100*]. Two years of attempts by the king to prevent the parlement from interfering culminated, at the end of 1751, in a judicial strike even more complete and determined than that of 1732. As then, led by a core of Jansenist diehards, the bar suspended its service too. Forced to resume their functions by direct royal orders, the magistrates found themselves bombarded throughout 1752 by appeals against refusal of sacraments. In the 14 months down to May 1753 there were no less than 22 cases, coming from six different dioceses [69: *ch. 4*]. When the archbishop defiantly refused to recognise the parlement's jurisdiction, it impounded his temporalities. Sooner or later, as under Fleury, most of these cases were evoked; but not before the parlement had formally forbidden parish priests to withhold the sacraments. Early in 1753, when this injunction was overridden in its turn, the parlement resolved to send the king new remonstrances. Unprecedentedly, their drafting was entrusted to a clutch of notoriously Jansenist magistrates, who took advice from sympathetic advocates and canonists. The result was the so-called 'grand remonstrances' which catalogued in massive detail the excesses of episcopal pretensions and went on to claim an exalted role for the parlement in the preservation of the fundamental laws of the kingdom. When Louis XV refused to receive such a sustained indictment of government policy and practices since 1713, the parlement allowed the remonstrances to be published and once again suspended service. In return, the king exiled the magistrates and replaced the parlement by a special 'royal chamber'.

More serious than the crisis of 1732, the confrontation that began in May 1753 lasted 15 months. In the publication of the remonstrances and the claims they made, and in the unprecedented support given to the parlement by its provincial counterparts, the confrontation marked a new direction in relations between the king and his courts. It also signalled a clear evolution in the character of Jansenism. Laymen and lawyers were now taking the lead permanently. Whether in drafting remonstrances or in deluging the public with supportive pamphlets, the advocates of the Paris bar were now more influential than the lingering remnants of the appellant clergy [46: *115–22*], or even the elusive *Nouvelles Ecclésiastiques*. A younger generation of activists was also emerging. The crisis of the 1750s brought to the fore the organisational and polemical talents of Louis Adrien

Le Paige (1712–1802), whose *Historical Letters on the essential functions of the Parlement* (1753–4) provided an arsenal of constitutional arguments for a generation of judicial resistance to the pretensions of authority. Appointed bailiff of the Temple, a jurisdictional peculiar in the gift of the prince de Conti, in 1756, he made it a headquarters of Jansenist polemical activity until the Revolution. The library he built up, now owned by the Friends of Port Royal, is still the most important collection of sources for the entire history of Jansenism.

For all its new departures, the crisis of 1753–4 eventually ended in time-honoured ways. The magistrates were recalled with ambiguous words, studiously avoiding the original issues, and the king dictated a new law of silence which most magistrates were eager to accept and implement. Resistance now came from the devout clergy, who continued to refuse sacraments to dying suspects with the open support of their archbishop. Now it was his turn to be exiled for disobedience. The king used his absence to open negotiations with Rome to secure the banning of confession notes.

Benedict XIV (1740–58), the greatest pope of the eighteenth century, was polished, educated and broad-minded. He found the passions aroused in France hard to understand. But once assured that his intervention was the key to stilling them, he readily agreed. By the brief *Ex Omnibus* (October 1756) he forbade denial of the last sacraments to anyone except 'public and notorious sinners'. Those who rejected *Unigenitus* certainly endangered their own salvation, but that was no ground for denying them life's last consolation.

To draw papal authority into a French domestic dispute, especially in a field littered with the debris of previous forays, was to risk reopening all the familiar jurisdictional problems. In a context of bitter ministerial rivalries at Versailles, and renewal of warfare in Europe and overseas, it presented the small, vocal kernel of Jansenist magistrates with a new opportunity to denounce threats to the parlement's authority [72: *ch. 5*]. When in December 1756 *Ex Omnibus* and a number of organisational measures were imposed in a *lit de justice*, these militants were able to persuade their colleagues to resign in protest. It was into this arena of recurrent crisis that, in January 1757, an element absent from French politics for a century and a half was suddenly thrown – regicide.

[iii] Damiens

Louis XV was only slightly wounded when Robert François Damiens, a jobbing domestic servant, stabbed him with a penknife; but the shock to the whole political system was profound [79]. After his conviction, the insistence on executing Damiens in exactly the same grisly manner as regicides of previous centuries showed what deep-rooted anxieties his lone deed had stirred. Those who interrogated him discovered something yet more disturbing: he appeared to have been influenced by talk critical of the king, overheard among magistrates whom he served. He seems to have thought he was punishing the king for supporting the tyranny of the bishops and clergy 'who ought not to refuse the sacraments to people who live well' [79: 45]. A crackdown on popular loose talk in the aftermath of the crisis revealed that Damiens's attitudes were widely shared. All sides in the ongoing political conflicts rushed to disclaim responsibility and apportion it elsewhere. Magistrates protested their horror at the act. The archbishop blamed the parlement and Jansenism for spreading insubordination. Jansenists and their sympathisers, noting that Damiens had once worked for Jesuits, instinctively cast suspicion on their traditional enemies, never free of old regicidal associations. When, the next year, news came of an alleged Jesuit attempt to assassinate the king of Portugal, these suspicions gained credibility. And when, on this pretext, the Society was expelled from Portugal and all her extensive overseas dominions, Jansenist minds were opened to the possibility of a more serious revenge, even in France.

Damiens's intervention came close to defusing the political crisis, when conscience-stricken magistrates in the parlement offered to resume their service unconditionally. But the king, who recovered within days, was too angry to accept this gesture. He even began moves to dispossess the leaders of the December defiance of their offices. This so inflamed matters that, despite the government's dire financial needs, it took many months, and the fall of a number of leading ministers and magistrates, to negotiate a resumption of service. But as part of that settlement, *Ex Omnibus* was at last registered.

It looked like one more solution obtained through studied ambiguity. The deeper but decisive significance unfolded more slowly in everyday practice [79: *154–63*]. The law of silence and the pope's brief were to be enforced not by the crown, but by the parlement. Evocations were abandoned, and protests by bishops now went unheeded. Those who persisted in trying to impose conformity by refusing sacraments faced exile, or were forced into resignation. Suspected Jansenists were allowed to die in peace and be buried in hallowed ground. Having relied on the bishops ever since 1695 to enforce uniformity, the crown now abandoned them. They had become the source of the problem rather than its solution. Under the duke de Choiseul, who came to power in 1758, religious policy was to be enforced in co-operation with the parlement, and Beaumont and his dwindling band of episcopal allies were reduced to relative impotence. Their last hope now lay in the eventual succession to the throne of the dauphin, who had become the focus of a devout party at Court. But he died in 1765. And by then Jansenism was well on the way to its last, though greatest, triumph: the destruction of the Jesuits.

No doubt, both for devout bishops and dying dissidents, the basic issue in the dispute over refusal of sacraments remained the content of *Unigenitus*. But there were far wider dimensions. Denial of the right of public authorities to impose terms on the dying, it has been argued, marks an important cultural shift towards the 'privatisation' of death [23]. Less plausibly, the volume of popular criticism directed at a king who flaunted an immoral life even as he supported hardline bishops has been seen as evidence of the incipient 'desacralisation' of the monarchy in the eyes of its subjects [52, 64, 79]. The authority of the monarch was certainly challenged in intellectual terms during the jurisdictional controversies produced by the quarrel. The grand remonstrances of 1753, largely drafted by Jansenists, were a sustained indictment of Louis XV's failure to maintain his authority against papal encroachments. Le Paige's contemporaneous *Historical Letters*, meanwhile, advanced claims for the parlement scarcely heard before – that it was coeval with the monarchy, that in some sense it represented the king's subjects, and that all the parlements were simply 'classes' of a single body empowered when appropriate to defy him. Other pamphlets by Jansenist lawyers such as Claude Mey (1712–96) or Gabriel Nicolas Maultrot (1714–1803), both now beginning a lifetime of polemical writings, broached parallel themes. None denied the king's authority. The professed aim of most, indeed, was to protect it against the claims of Rome. But when the king could not be trusted to uphold his own jurisdiction, his subjects must be able to do it for him, and institutions must prevent him from betraying his trust. Thus the parlements, primarily concerned as always to enforce the law and uphold their own jurisdiction, were elevated beyond their traditional role of defenders of the Gallican liberties, into the voice of the nation; and Jansenism can be seen as one of the sources of the idea of representative government in France, fully three decades before it came to fruition.

7 Wider Jansenism

Jansenism was never solely confined to France. The issues from which it arose were universal, and were first ventilated in the Netherlands. And insofar as much of what was called Jansenistic was a figment of Jesuit suspicions, Jansenists were likely to be identified wherever the Jesuits were established.

The best example was in Spain. The Iberian peninsula, the source of many of the writings held in most horror by Jansenists, was scarcely touched by the controversies they provoked in France. It was only in 1747, when the Spanish Inquisition published a new index of prohibited books with a special appendix of Jansenist works drawn up by Jesuits, that the Spanish Dominicans and Augustinians found that some of their most revered writers stood accused of this heresy [84, 95]. A horrified Benedict XIV correctly predicted that this would create Jansenists where none had existed, but he was precluded from intervening by the powerful anti-papal traditions of the Spanish monarchy. So that when, in 1759, a king intent on strengthening those traditions yet further came to the throne in the person of Charles III, he found willing collaborators among those whom the Jesuits had attacked (see below, pp. 73–4).

[i] Louvain eclipsed

Spain's isolation from the classic Jansenist controversies of the seventeenth century was all the more surprising given that the southern Netherlands were a Spanish dominion. Despite the constant sniping of the Jesuits, the university of Louvain, where most of the Belgian clergy were educated, remained loyal to the legacy of Baius and Jansen. It was only after Arnauld, Quesnel and other victims of Louis XIV's persecution sought refuge across the frontier in the 1680s that the Spanish authorities became alarmed. In 1692 a first attempt was made to impose the formulary of Alexander VII. As in France, all this did was to focus contention on the authority of the pope, and in 1700 the leading canon jurist of Louvain, Bernard Van Espen (1646–1728), produced *Jus Ecclesiasticum Universum*, a massive treatise which argued that the Church was fundamentally conciliar. The pope was merely first among equal bishops, and they in turn ought to be subordinate to the civil power. What could happen when they were, however, was demonstrated after a French claimant took the throne of Spain in 1700. The Jansenist exiles were arrested or fled to Holland, and only the fact that the Netherlands were a war zone impeded further decisive action.

The transfer of these territories to the House of Austria in 1713 coincided with the great reorientation in Jansenism brought about by *Unigenitus*. Appellants now found Van Espen's work a major inspiration, but he himself fell victim to orthodox persecution. It took 17 years to secure Louvain's unequivocal acceptance of the bull. The university's first reaction was to accept with reservations – a classic Jansenist ploy. But from 1718 a steady purge (which included refusal of sacraments) gradually removed appellants and equivocators alike. Van Espen lost his chair, and died in exile at Utrecht. By the 1740s Jansenism in its birthplace was dead; and when, half a century later, the emperor Joseph II attempted to introduce an element of what Jansenism had become in the form of a new vernacular breviary, Belgian seminarians would lead a revolt in defence of orthodoxy. The influence of the old Louvain was prolonged, however, by those who had been educated there in Van Espen's time. They included the Dutch doctor Gerard Van Swieten (1700–72) who, as an adviser to the empress Maria Theresia, consistently undermined the influence of the Jesuits in the Habsburg domains; and as part of that policy allowed the circulation of many Jansenist writings. Among the latter were the works of Nikolaus von Hontheim, called Febronius (1701–90), who had studied under Van Espen and introduced his Erastian, anti-papal doctrines into Germany (see below, pp. 74–5).

[ii] Dutch separatism

The decline of Louvain as the capital of non-French Jansenism was offset by the rise of Utrecht. Dutch Catholics, unlike most of their co-religionists, were not an established church. Accordingly, the secular arm did not appoint bishops. The clergy elected a vicar apostolic whom the pope routinely invested as titular archbishop. It was a pattern close to the clerical democracy advocated by Richer and commended by Quesnel; and informal links were established as early as the 1660s, through the ubiquitous Oratorians, between this relatively free Catholic community and Port Royal. Thus Holland was an obvious refuge for Arnauld and Quesnel and the presence of such figures left a permanent impression. And so when, after election and confirmation in 1686, archbishop Pieter Codde (1648–1710) was ordered by Rome to accept the formulary of Alexander VII, he and most of the other clergy refused. Except for a few Jesuits, reported a leading Oratorian in 1692, every Dutch priest was a Jansenist [12: *202*]. Rome suspended Codde, and dissolved the Utrecht chapter which had elected and still supported him, brushing aside legal objections provided by Van Espen. When the pope refused to provide a successor after Codde's death, the Jansenist exiles arranged for sympathetic French bishops to continue ordaining clergy. Naturally the rift with Rome was exacerbated by *Unigenitus*; and in 1724 a formal break occurred when a newly elected archbishop denied sanction by the pope received the apostolic succession from a renegade missionary bishop. A schismatical church now existed in Holland, and Rome always rejected its conditional offers of reconciliation. Though never more than a few thousand strong, with scarcely more than 200 clergy, it pursued ideals that won it many more admirers abroad: clerical election, a vernacular Bible and liturgy, an abstemious reverence for the sacraments, worship in surroundings of simple austerity. When ordinands could no longer be reliably educated at Louvain, clerical training was provided by seminaries at Rijnwyk and Amersfoort. From here, printing presses bombarded French-speaking Europe with reprints of classic Jansenist texts, histories of Port Royal and its heroes and heroines, as well as contemporary sectarian polemics. The libraries of the Dutch Old Catholics (some of whom still trace their history to the schism of 1724) remain a major source of documentation on every aspect of Jansenism, particularly the role of refugees who died at Utrecht. These included characters as diverse as Van Espen, the driest of lawyers, and the figurist exegete Etemare who, after a life spent predicting wonders, expired just three years before an event that he had not predicted: the destruction of the Jesuits.

[iii] The Jesuits destroyed, 1750–73

Jansenists were only one element in the complexities that led to the destruction of their oldest enemies. And they were certainly not alone in supporting and applauding the deed, since the Jesuits had many other enemies. Nevertheless, at certain moments the role played by Jansenists was crucial.

First, they needed to be convinced that destruction was a real possibility. Nobody thought so before 1750: Jesuits educated most of the elites of the Catholic world and confessed most of its kings. They did not win all their battles, as condemnation in the Chinese rites controversy, finally upheld by Benedict XIV in 1743, showed; but their basic position within Catholic christendom seemed impregnable. In the early 1750s, however, the Society came into direct conflict with the governments of Spain and Portugal by encouraging Indians in Paraguay to resist a rationalisation of frontiers between the two colonial powers. In the absence of clear secular authority in these remote regions, Jesuit missionaries had established a semi-independent protectorate which was threatened by the new agreement. The Indians fought for three years to remain separate, with open Jesuit support. Portuguese Jesuits even depicted the great Lisbon earthquake of 1755 as divine punishment for the kingdom's sins. In reply, the chief minister of king Joseph I, Pombal, accused the Jesuits of enslaving the Indians, and in 1757 expelled them from the Court. When in September 1758 a shot was fired at the king in obscure circumstances, the opportunity to implicate a group once notorious for countenancing regicide was irresistible. Recent events in France were invoked to clinch the case. And so, ignoring papal protests, in September 1759, Pombal expelled the Jesuits from Portugal and all her dominions – a not insubstantial proportion of the Catholic world.

Jansenists had nothing to do with any of this, but the agony of their old enemies galvanised them. In France, the arrival of the sensational news provoked immediate speculation about prospects for a similar onslaught. A new chief minister, Choiseul, let it be known that he would not impede an attack, and throughout 1760 Jansenist magistrates and publicists combined to remind the public of the iniquities of the Jesuits' historical record. What they lacked was an occasion for action; and it was provided by the Jesuits themselves in 1761 when they appealed to the parlement of Paris against a judgement of the consular (or commercial) court of Marseille . The hostilities of the Seven Years' War had occasioned the collapse of a wide-ranging trading network operated from the French island of Martinique by a free-wheeling Jesuit called Lavalette [77]. A series of lawsuits ensued, revolving around whether the French Jesuits as a whole were collectively liable for the losses incurred by Lavalette's trading partners. But when the issue came to a hearing it was rapidly widened into a judicial review of the whole status of the Society of Jesus in France. This approach was clearly concerted between the defendants' lawyer, the well-organised clique of Jansenist magistrates in the parlement, and the ubiquitous hidden hand behind Jansenist strategy, Le Paige. Thus, by the time the appeal went against the Society, a veteran practitioner of jurisdictional Jansenism, the clerical counsellor Chauvelin (1716–70), had denounced the Jesuits as an organisation incompatible with the laws of the kingdom (17 April 1761).

So began the ultimate triumph of the Friends of the Truth. The Jesuits were required to deposit a copy of their 'constitutions' for examination by the court. They were scarcely the secret document alleged by Chauvelin and the results were entirely predictable. The apparently blind obedience vowed by all Jesuits to their general and to the pope was found to contradict a subject's duty to the king. But the magistrates who made the examination were not Jansenists and their conclusions fell far short of the latter's hopes, merely suggesting that a new basis for Jesuit operations in France should be negotiated with the pope. The Jansenist attack was now switched to the perniciousness of the doctrines, regicide included, presumably poured daily into the innocent ears of the young in Jesuit colleges. And had the Society as it actually functioned ever been legally admitted to the kingdom? These were lawyers' arguments, but the parlement was a forum where they counted. And so on 6 August 1761, the parlement evaded a last-minute attempt to evoke the whole issue to the council and declared that the Jesuits had never had any legal existence in France. Consequently, French subjects were now forbidden to take Jesuit vows; and all the Society's colleges were ordered to close within a year.

The Jansenist magistrates who had orchestrated this outcome were well aware that such a delay left a loophole for counter-measures. They also knew that the parlement only exercised jurisdiction over a third of the kingdom. They now sought to make the momentum of their campaign unstoppable by urging the other parlements to follow the example. Le Paige used a network of provincial contacts (not all of them Jansenists) to spur the other sovereign courts into action. By the end of 1763 all but three had condemned the Jesuits in similar terms. Unsupported either by prior agreement with Rome or ministerial unanimity, half-hearted government attempts to stem the tide were swept aside. Only the devout party, the bishops, and a few Jesuit pamphleteers protested. By 1764 the Jesuit order had ceased to exist as a recognisable entity in most of France. A royal declaration of November confirmed the disbanding of the Society and the closure of all its schools.

Hitherto, most of Jansenism's history had been a catalogue of heroic failure or persistence against the odds; but this was a clear and total victory over the arch-enemy. Its cultural and educational consequences were enormous: 113 colleges were closed and those that reopened under new management were free to set their own syllabuses. The anti-clerical *philosophes* of the Enlightenment hailed the event as a great step forward for the emancipation of the human mind. Indeed, in 1765 d'Alembert claimed philosophic credit for their overthrow in his pamphlet *On the Destruction of the Jesuits in France, by a disinterested author*. Jansenists were outraged at this blatant attempt to attribute what they saw as the achievement of religious Truth to the influence of irreligion. In their eyes, there was little to choose between the laxities of Molinism and the free thought of Enlightenment. 'What is a true Jesuit', wrote Le Paige, 'if not a disguised philosophe, and what is a philosophe if not a disguised Jesuit?' [78: *218*]. But d'Alembert knew what he was doing. To allow the Jansenists the glory, he wrote to Voltaire, would be to boost the forces of intolerance. 'The Jesuits, amenable people so long as you do not declare yourself their enemy, are quite willing to let you think what you will; the Jansenists, as rude as they are ignorant, want you to think as they do; if they were the masters, they would exercise the most violent inquisition over writing, thinking, words and deeds.'

But the Jansenists were not the masters, and never would be. They were able to engineer the expulsion only because a dedicated handful of them were entrenched in the parlement and knew how to exploit its procedures and prejudices. Nor did Jansenists have much to do with the rest of the process that culminated in the final dissolution of the Society of Jesus by Pope Clement XIV in 1773. Once France had followed the Portuguese example, it became contagious. Shaken by riots against innovating economic and cultural policies in 1766, Charles III of Spain found it easy to blame an order already declared infamous by his two nearest neighbours. The Spanish Jesuits had already made too many enemies within the Iberian church by identifying Jansenism where none had existed, and they found few defenders when summarily ejected from Spain and her dominions in February 1767. The Bourbon states of Italy, still Spanish satellites, fell into line. Only Pope Clement XIII, who had resisted the anti-Jesuit tide throughout his pontificate (1758–69), stood between the Society and its complete dissolution. When he died the cardinals came under diplomatic pressure from all the Bourbon powers to elect a successor who would finish the job. Clement XIV (1769–74) took four years to do so, but he seems to have realised from the start that he had little alternative. The bull *Dominus ac Redemptor* (1773) made him a hero to the enlightened, to Protestants – and to Jansenists. In their euphoria, they did not notice that the dissolution of the most powerful order in the Church was an act of papal authority far more extensive than any against which their predecessors had appealed in the previous 130 years.

[iv] Febronianism

The sovereign with the least enthusiasm for the destruction of the Jesuits was the empress Maria Theresia. She merely stood by and let it happen. Nevertheless, the preceding two decades had seen a steady erosion of Jesuit power in the Habsburg lands. Adverse publicity from Portugal, France and Spain no doubt played its part, but so did the influence of Jansenist works of piety, known to the empress herself and recommended by her in the education of her son [85: *65–6, 441–4*]. A number of Jansenist classics, in fact, were printed at a private press in Bohemia by Count F. A. von Sporck [83: *146–7*] who had travelled in France in the 1690s and returned determined to spread the fame of Port Royal and its doctrines. Until prevented in 1733, he bombarded German-speaking lands with translations of Arnauld, Nicole, Pascal and other writers of the heroic age. German interest in Jansenistic ideas was also spread by the very vigour with which the Jesuits who ran the German College in Rome tried to rebut them. Among the German clergy who passed through Rome, after completing his studies at Louvain, was Febronius.

His book, *On the State of the Church and the lawful Power of the Pope, written to reunite Christians who differ in religion*, was published in 1763, after a lifetime's study. By then Febronius was suffragan bishop of Trier, and he had concluded that the key to reuniting the divided Christians of Germany was to diminish the power of the pope and govern the Church through regular general councils of bishops. Such basically Gallican ideas had often been espoused by Jansenists at odds with papal authority; although invoking them to bring Protestants back into the Church was a new departure. Their Erastian tone was also welcome to rulers eager to subject the Church in their dominions to closer control. Papal attempts to unmask the pseudonymous author lent the book notoriety, only increased when it was revealed that he was a bishop. His book was placed on the *Index*, but far from banning it, the governments of the Habsburg lands, Spain and Portugal, prescribed it for use in universities. When, under papal pressure in 1779, Febronius publicly retracted, the empress refused to allow the retraction to be published. Yet it was not so much monarchs as fellow bishops whom Febronius most directly inspired. When the elector of Bavaria, with no episcopal see on his own territory, persuaded the pope to appoint a nuncio to Munich, four prince archbishops met at Ems in 1786 to denounce papal interference in the Holy Roman Empire. The *Punctation* issued there claimed almost exclusive ecclesiastical jurisdiction within the empire for bishops and archbishops, reducing the pope to a distant primacy of honour. The emperor Joseph II, who had systematically diminished papal authority throughout his hereditary lands, despite a personal visit from Pius VI (1774–99) to Vienna in 1781–2, promised his support in securing Rome's acquiescence. But by the time the pope rejected the claims of the German bishops in 1789, the Church was facing far more serious problems than internal disputes over jurisdiction.

[v] Jansenism in Italy

The year 1786 saw the high point of Jansenism outside France. Not only did it witness the *Punctation* of Ems, it was also the year of the Synod of Pistoia, which brought Jansenism to an uneasy climax in Italy [86, 90].

The French Jansenists had always found protectors or allies at the papal curia, if only among those opposed to the influence of the Jesuits. Nevertheless, there was no significant Jansenism in Italy before *Unigenitus*. The claims made in the bull, however, were unacceptable to the traditionally anti-papal states of Venice and Sardinia, and this made them havens for dissidents. Under the conciliatory pontificate of Benedict XIV, Jansenist sympathisers proliferated in Rome itself. They met at the house of the Vatican librarian, Giovanni Bottari (1689–1775) [89: *ch. 12*] in a discussion group known as the *archetto*. Bottari was regarded by the Jansenists of France, not to mention Utrecht, as their key contact in Rome. The conflicts in France, the iniquities of the Jesuits, and how to strengthen true Christian commitment in Italy, were the main concerns of the *archetto*. They overlapped with those of earlier reformers such as the great antiquarian L. A. Muratori (1672–1750) who, though never professing sympathy with anything called Jansenist, consistently advocated a simplification of liturgical practices and improved clerical education to promote a more informed spiritual life among the laity. The destruction of the Jesuits, the death of Bottari, and the election of a less flexible pope in the person of Pius VI, brought the influence of the *archetto* to an end. Its ideals lived on and flourished in Tuscany, however, via a prelate who had frequented its discussions as a student in the late 1750s – Scipione de' Ricci (1741–1810).

Jesuit-educated (and related to the Jesuits' last general), Ricci reacted violently against their teachings in his teens. His return to Florence coincided with the arrival of the grand duke Peter Leopold, younger brother and ultimately successor of the emperor Joseph II. Leopold was as determined as his brother to assert secular power over the Church. But, again like his brother, he was also anxious to promote the spiritual well-being of his subjects by improving pastoral care. Ricci, well known for such ideals, was appointed bishop of Pistoia and Prato in 1780 and proved the dominant influence in Tuscan Church policy down to 1787. While enjoining his clergy to strict observance of the liturgical year, he urged them to discourage outward devotions such as veneration of images and relics, and to promote devout reading of sacred texts in the vernacular. The 'golden' spiritual guide he recommended was Quesnel's *Moral Reflexions* – an implicit rejection of *Unigenitus*. 'I think it necessary,' he wrote [86: 28], 'to flood the country with good books', by which he meant Jansenist classics imported from France and Utrecht, and translations of shorter polemics, printed at his own expense. Above all, he invited his lower clergy to share diocesan responsibilities, one of the few bishops ever to put Richerist principles into practice. He persuaded Leopold to authorise diocesan synods of parish priests to co-ordinate reforms with their bishops. These in turn would prepare for a general synod of the Tuscan Church. At the synod of Pistoia in 1786, Ricci set out to secure the co-operation of parish priests 'to combat and destroy... in the course of time the papal monarchy' [86: 55], which he described as diabolical and anti-Christian. At the same time he urged them to crack down on the superstitious excrescences of popular piety.

But when, having approved Ricci's proposals by overwhelming majorities, the clergy returned to their parishes and began to implement them, they came up against the basic weakness of nearly all Jansenism except perhaps that of Paris between about 1720 and 1760. It had no popular base. As the Tuscan bishops met in Florence to plan the general synod, riots broke out in Prato against rumoured plans to remove images from the cathedral. Churches were sacked, images reinstated, bonfires made of books introduced by Ricci. Troops were required to restore order. The furious grand duke abandoned plans for a national synod. Over his last three years in Florence he continued systematically to erode papal jurisdiction in the Tuscan Church, but when he left to succeed his brother in Vienna there were more riots against reform. This time Ricci was allowed to resign (1791); and three years later, immeasurably strengthened as the Church's defender against the godless attacks of the French Revolution, Pius VI felt emboldened to reassert traditional orthodoxy. In the bull *Auctorem Fidei* (1794) he condemned all that had been done at Pistoia. And although the Catholic rulers of Italy, Iberia and the Empire refused unanimously to accept the pope's renewed claims, the power of most of them was soon to be swept away by a French general who saw that secular rulers had more to gain than to lose by recognising Rome's authority.

Jansenism had begun as the creed of a tiny intellectual elite, and remained so when it spread beyond France. With the exception of the schismatical church of Utrecht, Jansenist ideas outside France were confined to a scattered, committed handful of canon jurists and moral reformers, nearly all of them clerics. United, both before and after 1773, by hostility to the Jesuits and the values and practices they epitomised, and relying for support on secular rulers, these late

Jansenists never won a steady lay following. During the few opportunities they had to impose their ideas, their puritanism aroused widespread popular hostility. Unlike their Jesuit opponents, they failed to see that the Church meant much more to the laity than the pursuit of religious truth and moral rectitude. Jansenists triumphant, determined (as d'Alembert saw) to make people think their way, directly challenged everyday habits, routines and prejudices which gave shape and meaning to the lives of innumerable ordinary, unintellectual Catholics. However imperfectly Christian, such ways had been tacitly condoned by the clergy. Clerics who now refused to condone them, in the name of doctrinal principle or priestly authority, found themselves deserted by those they sought to protect against error.

8 The Dissolution of Jansenism

With the destruction of the Jesuits, Jansenism fell to pieces. In one sense the Jesuits' creation, it lost its main uniting impulse when those who had first defined it were gone. For some time it sustained itself with the fear that they were not yet gone. Although members of the Society had been physically expelled from the Iberian kingdoms and the Bourbon states of Italy several years before 1773, former Jesuits were not required to leave France, and up to the last moment there was much speculation about their possible reinstatement. When in 1771, chancellor Maupeou struck at the parlements, the instruments of the Jesuits' downfall in the previous decade, the devout party hoped, and Jansenists feared, that the moment had come. Events in Rome had gone too far for that; but even after *Dominus ac Redemptor*, under a new pope who clearly regretted what his predecessor had done, paranoid suspicions continued that the Jesuits were still pursuing their nefarious ends under cover. Eventually, in 1814, the Society would indeed be reinstated. By then its dissolution looked in retrospect like the papacy's first fateful surrender to the irreligious onslaught which had culminated in the French Revolution. Yet at the time, many in the Church had regarded it as an opportunity to renew primitive Christian values the better to face the challenges of increasingly godless times.

[i] Against atheism

An unfailing barometer of what mattered to Jansenists in the eighteenth century is the content of the *Nouvelles Ecclésiastiques*. Up to the 1770s this most successful clandestine journal of the century was full of the iniquities of the pope, the bishops, and of course the Jesuits. Papal and episcopal pretensions continued to preoccupy it throughout the 1770s and 1780s, and the progress of Febronianism, the doings of the church of Utrecht, and Tuscan experiments in clerical democracy, received extensive attention. But now, increasingly, the voice of French Jansenism denounced the progress of free thought and irreligion [60: *72–5*]. The major works of the *philosophes* were unfailingly given hostile reviews: after all, in their commitment to human free will and capacity for self-improvement they were close to the Pelagians so execrated by St Augustine. Only the puritanical Rousseau, with his religious mind, contempt for worldly luxury and hostility to despotic power, struck some chords among late Jansenists [62: *81–100*].

The sense that religion generally was under attack led some Jansenists in unprecedented directions. The founders of Jansenism, for instance, had all been keen to distance themselves from the Protestants with whom their enemies tried to identify them. Arnauld had denounced Calvinists consistently, and thought it wrong for Catholic rulers to tolerate Protestantism. Across the Channel, it is true, some Anglicans had expressed hopes that common ground could be established with Catholics so clearly at odds with the pope; and at the time of the appeal against *Unigenitus* certain Jansenists did tentatively respond to feelers put out by the archbishop of Canterbury [4: *ch. 18*]. These efforts proved vain; but forty years later two generations of persecution had begun to persuade Jansenists of the virtues of religious toleration. Even as Febronius denounced papal pretensions as an impediment to reunion with the Protestants, leading Jansenist writers in France, such as Le Paige and Maultrot, were beginning to argue that French Protestants should be accorded toleration to worship as they liked. And when, in 1787, toleration was at last proclaimed, it was largely owing to the efforts of a leading Jansenist magistrate in the parlement of Paris, Robert de Saint-Vincent (1725–99) [62: *155–6*; 81: *341–4*].

In Lorraine, meanwhile (always a hotbed of provincial Jansenism through its network of Benedictine monasteries), an obscure young parish priest, Henri Grégoire (1750–1831), had begun to champion the civil rights of Jews. Later, after a career in the foreground of revolutionary politics, he would work for slave emancipation and wider racial equality [66].

[ii] Against despotism

Although Jansenism had always been sustained and protected at crucial points by sympathetic bishops, the vast majority of the

French episcopate had always been its most resolute persecutors. And even after the crisis of the late 1750s, when over-zealous prelates no longer received royal support, the damage, in Jansenist minds, was done. Bishops were tools of the pope, or of the Jesuits, whom they persistently tried to save. They were natural despots who needed to be restrained by their lower clergy. And so Richerism, so prominent in the 1650s, and in the prescriptions of Quesnel, now resurfaced in France as well as in Ricci's Tuscany. It thrived among the clergy of parishes whose tithes had been impropriated by monasteries, and who therefore lived on fixed stipends (*portions congrues*), constantly eroded by the inflation of the later eighteenth century. A further grievance was the under-representation of parish clergy on diocesan panels which apportioned levies to service the clergy's debts. From mid-century, unauthorised assemblies of parish clergy began to convene to air such grievances. When bishops tried to stop them, their authority was challenged in the courts, and Jansenist lawyers provided anti-episcopal briefs. *Rights of Rectors* (*Droits des Curés*), a polemic of 1776 by the crusading Dauphin priest Henri Reymond (1737–1820), used arguments drawn from classic Richerist texts. It was widely read, and spurred a surge in unauthorised clerical assemblies which, by the early 1780s, was being called a 'rectors' revolt' [73: *236–48*]. It was brought to an abrupt end in 1782 by a royal declaration forbidding all clerical assemblies without episcopal permission. Within a few years, however, the hopes then dashed were revived by the pre-revolutionary crisis. Lent focus by a series of anti-episcopal pamphlets by the indefatigable Maultrot, the complaints articulated during the rectors' revolt poured out in the *cahiers* of the clerical electors to the estates-general of 1789.

By then, Jansenist thinking had penetrated far beyond ecclesiastical affairs. Emerging originally in France from the efforts of the devout party under Louis XIII to make the state serve Catholic ends, Jansenism had always been profoundly political. Even the withdrawal from the world, propounded by the early solitaries, was seen as having important political implications. When, after 1680, state policy turned decisively towards persecution, leading Jansenist thinkers were forced to reflect more consistently on questions of political legitimacy [11]. Duguet's *Institution of a Prince*, written in 1699 but not published until 1739, after its author's death, rapidly became the key political handbook. A true Christian ruler, Duguet argued, should regard his subjects' service as his supreme duty. He should avoid unnecessary wars, take a close interest in all economic matters, and discourage usury. Above all, he should avoid dependence on self-serving ministers and listen to intermediary bodies like the parlements who expressed his people's grievances. The quarrels over refusal of sacraments later led a younger generation to place increasing emphasis on the courts which were doing so much to protect the Truth. Without the parlements, they recognised that the Jesuits could not have been dislodged.

When, therefore, in 1771, Maupeou attacked the bases of the parlements' power, Jansenists rallied to their defence. Some of the most influential pamphlets provoked by this great upheaval were the work of Jansenist lawyers, and distributed by Jansenist networks orchestrated by Le Paige. Despite suspicions that Maupeou was in league with the Jesuits, what was remarkable about this outpouring was how little religious issues were invoked. The questions of despotism, legitimate restraints on power, and how the people's voice could be made to prevail, were now central rather than implicit [62: *183–93*; 81: *251–81*]. The Jansenist party, wrote a commentator in 1771, had lost its true interest, and had transformed itself into the 'party of patriotism', opposed to the government. But writers like Le Paige, steeped though they were in the legalistic rhetoric of the parlements, were now losing their illusions about what traditional politics could achieve. The powers of the parlements, even legitimately exercised, were simply not enough to sustain what was increasingly depicted as a contract between the king and his people. And so Jansenists were among the first to call for genuinely representative institutions, elected assemblies which would spell the end of absolute monarchy.

It is true that the death of Louis XV in 1774 brought the downfall of Maupeou, and a restoration of the old parlements. Jansenists duly celebrated. But the reign of Louis XVI was to bring them little reassurance. Now they found their old champions, the parlements, alarmed by the radicalism unleashed against Maupeou, forming an almost unprecedented front with the bishops. After the death of the magistrate-hating archbishop Beaumont in 1781, prelates and magistrates co-operated in spurning the aspirations of the parish priests and resisting the granting of civil rights to God-fearing Protestants. So that by the time the old regime entered its terminal crisis, few Jansenists were prepared to defend any part of it; and they welcomed the opportunity to create a completely new order in Church and state after 1789.

[iii] For and against the Revolution

Almost from the start, those baffled by the French Revolution instinctively sought to explain it as the result of a plot. Most of them blamed godless philosophers or freemasons; but an influential strand of interpretation blamed Jansenism. According to this view, Jansenists had always been republicans and heretics, and the Revolution's break with Rome and overthrow of Louis XVI were the 'revenge of Port Royal' on the papacy and the House of Bourbon. Until recently, serious historians tended to dismiss this idea out of hand. Yet Jansenism in France had consistently been a reflex of opposition to authority and was often called republican. By the eighteenth century it had begun to rationalise its adversarial instincts; and the tendency of modern scholarship has been to emphasise the contribution of Jansenist theorists, jurists and pamphleteers to the elaboration of the ideology of national sovereignty and citizens' rights that emerged in 1788–9 [81: *ch. 4*].

At two crucial points in the Revolution, moreover, the influence of Jansenism was, if not decisive, certainly obvious. One was in the elections of the clergy in 1789, where parish priests had their revenge on the bishops for their suppression of the rectors' revolt. Although priestly Jansenists elected probably did not number more than 30 at the most, they did include Grégoire [75: *66*]. Many of the clerical *cahiers* which accompanied the election were full of Richerist rhetoric against the despotism of the bishops. This, and the relatively low number of bishops and regulars returned, gave other deputies a clear, though misleading, impression that there was a deeper mandate for reform among the clergy than actually existed. In fact, the most radical Jansenists had long been laymen. Even fewer of them secured election in 1789 than clerics, but they were all learned lawyers and canon jurists, obvious candidates for the ecclesiastical committee entrusted with reform of the Church. And so the civil constitution of the clergy which the committee produced in the spring of 1790, and which its Jansenist members fairly bludgeoned the Assembly into passing with the weight of their learning and oratorical stamina [75: *290–1*; 81: *353–60*], was shot through with their influence. Its hostility to the pope, subjection of bishops to election, and emphasis on the active role of the lay faithful, as well as a number of (now) lesser matters like the prohibition of formularies, were clearly of Jansenist inspiration. 'Far from doing any damage to religion,' one of this

83

group told the Assembly, 'your decrees will bring back its pristine purity; you will then find yourselves reborn as Christians of the evangelical era, Christians like the Apostles and their first disciples' [81: *356*].

Nothing of the sort happened. Denied the opportunity to consider the civil constitution through the eminently Gallican procedure of a national council, the French clergy were left with no way of ascertaining whether the Church accepted the changes – except through the pope. His rejection, though delayed, was unambiguous. The hesitations of a clergy torn between the sovereign nation and the sovereign pontiff led the Assembly to force them to choose by subjecting them to a secular formulary: the clerical oath of 1791. Grégoire was the very first to take it. Subsequently he accepted lay election as a bishop under the new procedures. So did Reymond. The *Nouvelles Ecclésiastiques* also welcomed the new order. But it soon became apparent that not all Jansenists approved the massive extension of secular control over Christ's church, especially at the hands of an Assembly which had earlier refused to declare Catholicism the national religion. Nor was there any precedent or obvious authority for the laity electing their clergy. On such issues Maultrot and Le Paige took opposite sides. And although oath-taking was high in the heartlands of the rectors' revolt, such as Reymond's south-east, there was no clear nation-wide correlation between acceptance and a previous history of Jansenism [74: *ch. 6*]. Ugly scenes of anti-clericalism marked the ceremonies whatever choice was made. The civil constitution, in fact, which owed so much to Jansenist influence, blew Jansenism apart. The oath made impossible an ambiguity on which it had always thrived – that it was perfectly possible to be a good Catholic and a good subject or citizen without any conflict of loyalty.

[iv] Death throes

The French Revolution proved to be the most serious challenge to the Catholic Church since the Reformation. Beside it, Jansenism seemed no more than a long-standing irritant. The eventual refusal of a majority of the French clergy to take the oath brought them under increasing persecution. It even tainted the 'constitutionals' who found the oath acceptable. Once war broke out, the suspicion of treason fell on all priests indiscriminately and by the end of 1793 public worship in the largest Catholic community in Europe had virtually disappeared under a wave of 'dechristianisation'. Not until 1801 did the French government again recognise the standing and authority of the papacy, and by then a decade of mortal conflict had reordered the Church's priorities.

Pius VI profited from the panic of his spiritual subjects to settle what now seemed less taxing issues for good. And so the bull *Auctorem Fidei*, elaborated while the Terror raged in France, was far more than a condemnation of the Synod of Pistoia (see above, p. 77). In anathematising what the synod had approved and recommended, it implicitly condemned much of what was now perceived as Jansenism. This bull was a new *Unigenitus*, condemning conciliarism, Erastianism (in the form of the Gallican articles of 1682), iconoclasm, vernacular worship, and all the writings commended by Ricci – including, of course, Quesnel's *Moral Reflexions*. But unlike *Unigenitus*, it provoked no appeals. Almost all those who still believed in these things were now outside the Church.

Although Catholic monarchs refused to publish such an unequivocal rejection of their jurisdiction claims, their priorities were changing fast as well. The execution of the Most Christian King of France was a terrible warning to all rulers. Church and king, they now realised, stood or fell together, and instead of quarrelling, should support each other's claims. When Pius VI died a French captive in 1799, his former capital proclaimed a Roman republic by godless Jacobins, the end of the papacy itself seemed at hand. For six months there was to be no pope. But when emperor Francis II, the son of Leopold, allowed a conclave to convene in Venice (traditionally the most anti-papal of Italian cities) to elect a successor, the way the world had changed was thrown into relief. And when even the new ruler of France welcomed the election of Pius VII, it was clear that the papacy had survived the worst of the revolutionary crisis.

This was the final blow to Jansenism. The excesses of dechristianisation in France proved transient. Religious practice soon resurfaced and proved impossible to eradicate again. And since so many non-juring clergy had been deported or killed, this looked like an unexpected opportunity for clergy who had taken the oath. Progressively abandoned between September 1794 and February 1795 by the state which had created it, but secure in its apostolic succession, the 'constitutional' church now proudly called itself Gallican. It was much like the church of Utrecht in doctrine and practice, and naturally Jansenism's last refuge. Its leading spirit was Grégoire, who showed where his sympathies lay by attempting to invite Ricci and clerics from Utrecht to a national council in 1797. During the agony of the papacy, a sustained effort was made to capture the religious revival; and it was a blow to these hopes when Napoleon Bonaparte opened negotiations with Pius VII for a formal restoration of the altars, and with them, the non-juring clergy. It is true that a further Gallican council was allowed to sit while the negotiations progressed, but this was only to force Rome's hand. Once a concordat was signed, it was summarily dissolved. The nomination of twelve Gallican bishops to the restored French episcopate disgusted Rome, but offered no other consolation to offset the massive return of the non-jurors. And, meanwhile, the powers conceded to the pope by the concordat far exceeded anything that old regime kings or bishops would have tolerated. In the expectation of a belated vindication, in 1801 Grégoire had produced an elegiac compilation called *The Ruins of Port Royal*. It proved an unexpectedly appropriate title, in the circumstances.

Jansenism – though not its more diffuse influence – was finished. Le Paige, its chief orchestrator in France since the 1750s, died in 1802. The next year, the *Nouvelles Ecclésiastiques*, now distributed from Utrecht, ceased publication. In 1805 Ricci submitted to *Auctorem Fidei*, though with equivocations and ambiguities worthy of Arnauld or Noailles. Only in Spain, where Jansenism arrived so late, did the monarch continue to support reforming bishops [84, 95]. But after the fall of the Bourbons in 1808, Spanish Jansenists, too, were left a shrinking handful, scarcely recognised by the isolated individuals outside the peninsula who still struggled to keep the memory of the Truth alive. No doubt they rejoiced when Napoleon annexed the papal states and subjected Pius VII to new humiliations at the hands of a puppet national council of the French Church. But this was the pope who, the moment his tormentor fell, celebrated the defeat of all his enemies in 1814 with the resurrection of the Jesuits.

[v] The significance of Jansenism

'One thing we cannot do', writes a most distinguished authority on eighteenth-century religion, 'is to define "jansenism"' [60: *254*].

86

Another, after a scholarly lifetime studying it, could only conclude that it was the opposite of anti-Jansenism [2: 5]. Its best-known historian denied its very existence, following Arnauld in calling it a phantom invented by the Jesuits [6]. Later writers have tended to avoid the problem by arguing that there was not one Jansenism but several [12: 9–13; 95: 49–61]. This begs the question of why so many have found a single name satisfactory.

Many difficulties are avoided if we look at Jansenism not as a body of doctrine but as a series of historical situations. Doctrines were involved, obviously, at every stage; and among those called Jansenists, and even that minority among them willing to be so called, the doctrinal diversity was wide and often inconsistent. What they had in common was not so much content as a tendency. Jansenism meant resistance to living authority in the Catholic Church. It was also confined to a limited period between the mid-sixteenth and the early nineteenth centuries. Although many of the ideas and reflexes called Jansenist, and most obviously those derived from St Augustine, were not unprecedented in the history of the Church, they only came together under Jansen's name in the 1640s, and were never traced back for much more than a century beforehand. Jansenism was specific to the post-Tridentine Catholic Church, a Church that had reorientated itself to meet and overcome the challenges of the Reformation. It withered away as a significant heterodox tendency when the challenge of Protestantism was supplanted by that of outright unbelief.

For all the repeated charges of crypto-Calvinism, Jansenist professions of loyalty to the Church must be accepted as sincere. In their commitment to the sacraments, to the priestly functions which alone could deliver them, to episcopacy (however diversely interpreted), and to the supremacy of the pope (unless expelled by him), they were indelibly Catholic. But equally, Jansenists claimed the right to criticise certain directions taken by the Church since the Council of Trent. They began by resisting lax modern theology in the name of the primitive fathers of the Church. Condemned for that, they resisted the papal and episcopal authority issuing the condemnations, and the Jesuits who had done so much to secure them. They also resisted anything that seemed designed to make the practice of religion easier or less demanding, whether in routine devotions, taking the sacraments for granted, or practices dependent on gratifying the eye, the ear, or the emotions. The Church should not be some form of theatre: religion was about the Word. That meant making it more accessible through translations of the scriptures and experiments with a vernacular liturgy. All these things the counter-reformation Church viewed with suspicion, initially as too Protestant, and subsequently as the thin end of a free-thinking wedge. The challenge of the French Revolution, echoing down through the nineteenth century, ensured that these suspicions triumphed. But in the twentieth, different challenges brought belated acceptability for some doctrines and practices once condemned as Jansenistic.

The Catholic Church, brought up to date in the 1960s not by infallible papal decisions but by a general council, now worships chiefly in the vernacular, and preferably in surroundings of scarcely adorned simplicity. Priests celebrate the sacraments with maximum lay participation and encourage the faithful to read their Bibles. And although a theology of free will, once called Molinism, is dominant in a Church which needs to win and keep its faithful, the Jesuits are once more under suspicion in many quarters for the flexibility of their approach to the Church's problems.

Because, under the old regime, church and state were so tightly integrated, defiance of one authority inevitably embroiled the other. For much of its history Jansenism flourished by playing them off against each other. The consequence was normally for Jansenists to find themselves vaunting secular authority against ecclesiastical, whether as Gallicans in France, Regalists in Spain or Italy, Febronians or Josephists in Germany and Austria. After the original theological divisions, in fact, much of Jansenism's history must be written in terms of jurisdictional conflict, an affair of lawyers rather than divines. In France, however, the fight was three-cornered, and the persistent if intermittent determination of absolute monarchs and their ministers to impose doctrinal orthodoxy led Jansenists to seek protection from the courts of law. In so doing, they elaborated rationales of resistance to the king as well as to the pope, and these arguments continued to be deployed even when sacraments were no longer refused to opponents of *Unigenitus*, and the king stopped supporting the bishops and the Jesuits. By the 1770s, the Jansenist publicity machine was the main source of opposition to the ambitions of absolute monarchy, at a time when the *philosophes* of the Enlightenment still wavered uncertainly about where the best interests of the nation lay. Claims that the Jansenist quarrels of the mid-eighteenth century led to a desacralisation of the French monarchy in the public mind are unconvincing. But what those advancing them have shown is that the arguments of Jansenists against established authority did as much as those of the more familiar writers of the Enlightenment to sap the intellectual foundations of the old regime; and that a notoriously anti-religious revolution could nevertheless have important religious origins [64, 81].

Much was once made of Jansenism's social significance. A whole generation of mid-twentieth-century historians felt compelled to take seriously Lucien Goldmann's Marxist argument that the Jansenism of the seventeenth century, at least, was the class-response of office-holders facing social decline. But it was pointed out from the start that Goldmann's views owed more to theory than to evidence, and that even such evidence as he relied on was far from unambiguous [27, 28]. Nevertheless there *was* a social dimension to Jansenism. It was almost always confined to a narrow intellectual elite. With the exception (which many of its leaders found distasteful and disturbing) of the popular support given to living, dead or dying martyrs to the Truth in Paris under Louis XV, Jansenists were all highly educated. They came from backgrounds affluent enough to afford that expensive process. Jansenism always flourished in universities and centres of learning, whether theological or legal. From there it seeped into the professional life of the Church and the law. Never exclusively clerical, it looked from the start to the educated laity for support, and by the end perhaps a majority of its leading spokesmen were not priests. Those who were, however, and gained the support of the secular arm to impose their beliefs, soon learned how limited was the appeal of their austere intellectualism to the Catholic masses. The Word was not enough for people who could not read it; and not until an age of mass literacy could the Church think of embracing Jansenism's pastoral values to retain, rather than alienate, the faithful.

Never a popular movement, Jansenism remains a symptom of an important long-term socio-cultural development: the emergence over early modern times of an ever-expanding educated community, less and less content to live by the dictates of authority (though still ready enough to impose its own). It was not, of course, the only such symptom. In a sense the Protestant Reformation was one, too. That was one meaning of the priesthood of all believers. Later, the Enlightenment would appeal directly to the critical common sense of an educated laity. Jansenists consistently invoked ancient authority against modern. But in their refusal to allow those in charge of their Church to dictate everything they should believe, they were authentic harbingers of the modern world. And the significance of Jansenism was to show that the Catholic Church itself could no longer expect unconditional obedience, even from those with no desire to abandon it.

Appendix

1 The Five Propositions

i. Some of God's commandments are impossible for the Just to obey, because they lack the grace to make it possible.

ii. In the state of fallen nature, there can be no resistance to interior grace.

iii. For merit and demerit after the fall, it is not necessary to have freedom from necessity, it is enough to have freedom from constraint.

iv. The Semi-Pelagians admitted the necessity of an interior and prevenient grace, even for the beginning of faith, but they were heretical in holding that it might be either resisted or obeyed.

v. It is a Semi-Pelagian error to say that Christ died for all men.

2 The Formulary of Alexander VII, 1665

I . . . submit myself to the apostolic constitution of Innocent X dated May 31st 1653, and to the constitution of Alexander VII, dated October 16th 1656, the supreme pontiffs: And I do with a sincere mind reject and condemn the five propositions, taken out of Cornelius Jansen's book, called *Augustinus*, and in the sense intended by the same author, as the apostolic see has by the said constitution condemned them.

Glossary

absolution   forgiveness of sins

apostolic succession the transfer of spiritual powers to bishops from previous bishops going back to St Peter

appel comme d'abus   appeal to a secular court against decisions of ecclesiastical authorities in pre-revolutionary France

Arminianism   Protestant doctrine of achieving salvation through free exercise of the will; named after the Dutch theologian Arminius (1560–1609)

attrition   sorrow for sins resulting from fear of God's anger. *See also* **contrition**

benefice   a position in the church

breviary   prayer-book containing the texts of services

bull   the highest and most formal papal pronouncement on a matter of doctrine

cahiers   lists of grievances drawn up by electors to the Estates-General

Calvinism   extreme form of Protestantism deriving from Jean Calvin (1509–64), emphasising predestination and the impotence of human will without the grant of God's grace. *See also* **grace, predestination**

casuistry
the practice of assessing and prescribing moral conduct in the light of precedents. *See also* **Laxism**

conciliarism
doctrine that councils have more authority in the church than the pope

confession
the confession of sins to a priest, who confers absolution, sometimes conditionally on performance of a penance. *See also* **absolution**

contrition
sincere sorrow for sin resulting from love for God. *See also* **attrition**

curia
the Court of Rome, the seat of papal government

Erastianism
doctrine of state authority over the Church

evocation
removal of a case from ordinary to a higher extra-ordinary jurisdiction.

Gallicanism
doctrine of the jurisdictional independence of the French Church from the temporal (but not spiritual) authority of the pope

grace
assistance granted by God to humans in order to achieve the eternal life of salvation

index
offical list of books which Catholics are not permitted to read or possess

incumbent
holder of a benefice

Inquisition	supreme papal court in cases of heresy
last rites	the conferment of the sacrament of extreme unction on the dying. *See also* **sacraments**
Laxism	an extreme form of casuistry, used to justify a degree of moral flexibility. *See also* **casuistry**
lit de justice	special sitting of a parlement held in the royal presence in order to override judicial resistance to legislation. *See also* **parlement**
Molinism	Catholic doctrine of achieving salvation through free exercise of the will, deriving from the Jesuit, Luis de Molina (1535–1600)
peculiar	area exempt from normal jurisdiction
Pelagianism	fifth-century heresy of Pelagius, emphasising the role of human free will in achieving salvation
philosophe	popularising writer of the Enlightenment
portion congrue	the proportion of impropriated tithes allowed to an incumbent by the tithe owner. *See also* **tithes**
parlement	a sovereign, *i.e.* final court of appeal in pre-revolutionary France
predestination	doctrine that God's creatures are doomed either to damnation or salvation, irrespective of their own deeds or efforts

regale, **regalian rights**   king's right to enjoy the temporalities of a vacant bishopric. *See also* **temporalities**

sacraments   the means by which God's grace is conferred. There are seven: baptism, confirmation, communion, penance, extreme unction, holy orders, and matrimony. *See also* **grace**.

temporalities   income attaching to an ecclesiastical benefice

tithe   notional tenth of the goods of the laity assigned for the upkeep of a parish priest. **Impropriation** of tithes is the acquisition of the right to collect them by someone other than the incumbent. *See also* **benefice**, **incumbent**.

Bibliography

The following list of books is in no way exhaustive. The bibliography of Jansenism is truly vast, expanding all the time, and could never be adequately covered in a work of this scope. This list, therefore, is restricted to works in English and the more important contributions in French and other languages from which this book has been written. Fortunately most of these books have full bibliographies of their own from which most aspects of the subject can be followed up in greater depth.

General works

[1] H. Brémond, *Histoire Littéraire du sentiment religieux en France* (Paris, 1916–33), 11 vols. Volume 4 of this classic survey attempts to reclaim the leading Jansenists for orthodoxy. Unfortunately, only the first three volumes have been translated into English, but vol. 3 contains an important discussion of Bérulle.

[2] L. Ceyssens, 'Le Jansénisme. Considérations historiques préliminaires à sa notion', in *Nuove ricerche storiche sul Giansenismo* (Analecta Gregoriana, lxxi) (Rome, 1954). Cautious thoughts of a leading twentieth century authority, introducing a collection of wider significance.

[3] P. Chaunu, 'Jansénisme et frontière de catholicité', *Revue Historique*, 227 (1962), pp. 115–38. A review of [10], arguing that Jansenism was at its most vigorous near to regions of Protestantism.

[4] R. Clark, *Strangers and Soujourners at Port Royal* (Cambridge, 1932). Surveys connections between the British Isles and French and Dutch Jansenism.

[5]   L. Cognet, *Le Jansénisme* (Paris, 1961). A masterpiece of succinct summary in the *Que sais-je?* series.

[6]   A. Gazier, *Histoire générale du mouvement janséniste depuis ses origines jusqu'à nos jours* (Paris, 1923), 2 vols. Unashamedly partisan history by the custodian of the most important Jansenist archive, the library of the Friends of Port Royal. Invaluable for its range, but to be read with caution.

[7]   C. L. Maire, 'Port Royal: the Jansenist schism', in P. Nora (ed.) *Realms of Memory. The construction of the French Past. Vol. I Conflicts and Divisions* (New York, 1966), pp. 301–51. A reflective and well-informed essay precedes the most up-to-date survey in English of the historiography of Jansenism.

[8]   J. Plainemaison, 'Qu'est-ce que le Jansénisme?', *Revue Historique*, 553 (1985), pp. 117–30.

[9]   C. A. Sainte-Beuve, *Port Royal* (Paris, 1840). There are many later editions of this classic nineteenth-century survey of seventeenth-century Jansenism, the greatest work ever written on the whole subject. Though superseded at innumerable points by later research, it is still the indispensable starting point. The edition used in writing the present book was that of 1908 in seven volumes.

[10]   R. Taveneaux, *Le Jansénisme en Lorraine, 1640–1789* (Paris, 1960). Definitive survey of Jansenism in a frontier province, pioneering the study of non-Parisian activity.

[11]   R. Taveneaux, *Jansénisme et politique* (Paris, 1965). Useful collection of extracts from key texts, with commentary.

[12]   R. Taveneaux, *La vie quotidienne des Jansénistes* (Paris, 1973). Useful sketches of aspects of Jansenist practice.

Background

[13]   J. Bossy, 'The social history of confession in the age of the reformation', *Transactions of the Royal Historical Society*, 5th series, 25 (1975).

[14]   P. Brown, *Augustine of Hippo. A biography* (London, 1967). Readable introduction to the saint, his outlook and his opponents.

[15]   L. Châtellier, *The Europe of the Devout. The Catholic Reformation and the formation of a new society* (Cambridge, 1989).

[16] N. S. Davidson, *The Counter-Reformation* (Oxford, 1987). Useful brief introduction.

[17] J. Delumeau, *Catholicism between Luther and Voltaire. A new view of the Counter-Reformation* (London, 1977). Translated from the French edition of 1971, contains a fine chapter anatomising the essentials of Jansenism.

[18] H. O. Evenett, *The Spirit of the Counter-Reformation* (ed. J. Bossy) (Cambridge, 1968). Stimulating reflections based on a lifetime's study.

[19] A. McGrath, *Reformation Thought. An introduction* (Oxford, 1988).

[20] R. Mousnier, *The Assassination of Henry IV. The tyrannicide problem and the consolidation of the French absolute monarchy in the early 17th century* (London, 1973). Translation of the French edition of 1964.

[21] E. Préclin and E. Jarry, *Les luttes politiques et doctrinales aux xviie et xviiie siècles* (Paris, 1956), 2 vols. (Vol. 19 of A. Fliche and V. Martin (eds), *Histoire de l'Eglise*). Essential general survey of the period of church history in which Jansenism flourished.

[22] E. Rapley, *The Dévotes. Women and church in seventeenth-century France* (Montreal, 1993).

[23] J. Delumeau, *L'aveu et le pardon. Les difficultés de confession, xiiie–xviiie siècles* (Paris, 1990).

[24] A. D. Wright, *The Counter-Reformation. Catholic Europe and the Non-Christian World* (London, 1982). Brilliant survey requiring slow and careful reading. Brimming with ideas.

Before Augustinus

[25] N. Abercrombie, *The Origins of Jansenism* (Oxford, 1936). Excellent on theological questions, but superseded on many historical matters by subsequent research. Contains (pp. 126–53) a full summary of the contents of Jansen's *Augustinus*.

[26] L. Cognet, *La Mère Angélique et son temps* (Paris, 1950–1), 2 vols.

[27] L. Goldmann, *The Hidden God. A study of tragic vision in the Pensées of Pascal and the tragedies of Racine* (London, 1964;

trans. of the 1955 Paris edition). Once-fashionable Marxist analysis of the links between Jansenism and the nobility of the robe. Now seems astonishingly crude and reductionist. The definitive critique is [28].

[28]   A. N. Hamscher, 'The *Parlement* of Paris and the social interpretation of early French Jansenism', *Catholic Historical Review*, 63 (July 1977), pp. 392–410. Definitive critique of [27].

[29]   J. Orcibal, *Les origines du Jansénisme vol. I Correspondance de Jansénius* (Louvain/Paris, 1947). Prints the key sources on Jansen's life.

[30]   J. Orcibal, *Les origines du Jansénisme vol. II Jean Duvergier de Hauranne, Abbé de Saint-Cyran, et son temps* (Paris, 1947). The authoritative source on Saint-Cyran.

[31]   E. Préclin, 'Edmond Richer', *Revue d'Histoire Moderne*, 29 (1930), pp. 242–69.

[32]   L. Willaert, *Les origines du Jansénisme dans les Pays-Bas catholiques* (Paris, 1948).

[33]   C. E. Williams, *The French Oratorians and Absolutism, 1611–1641* (New York /Bern, 1989).

Before *Unigenitus*

[34]   A. Adam, *Du mysticisme à la révolte. Les Jansénistes du xviie siècle* (Paris, 1968). Brilliantly readable account of the heroic age of Jansenism, by a historian of literature.

[35]   P. Blet, *Le clergé de France, Louis XIV et le Saint-Siège de 1665 à 1715* (Vatican City, 1989).

[36]   R. Briggs, 'The Catholic Puritans. Jansenists and Rigorists in France', in *Communities of Belief. Cultural and Social Tension in Early Modern France* (Oxford, 1989), pp. 339–63.

[37]   R. M. Golden, *The Godly Rebellion. Parisian curés and the religious Fronde, 1652–1662* (Chapel Hill, NC, 1981). Convincingly widens the known audience for early Jansenism.

[38]   J. M. Grès-Gayer, *Le Jansénisme en Sorbonne, 1643–1656* (Paris, 1996).

[39]   E. Jacques, *Les années d'exil d'Antoine Arnauld (1679–1694)* (Louvain, 1976).

[40]   P. Jansen, *Le Cardinal Mazarin et le mouvement janséniste français* (Paris, 1967).

[41]   R. A. Knox, *Enthusiasm. A chapter in the history of religion* (Oxford, 1950). Accessible but somewhat patronising essays by an English *monsignor.* Jansenism is covered in chs 9, 10, 16.

[42]   H. Lefebvre, *Pascal* (Paris, 1949–54), 2 vols.

[43]   A. Le Roy, *Un Janséniste en exil. Correspondance de Pasquier Quesnel* (Paris, 1900) 2 vols. Basic source on the founder of the second Jansenism.

[44]   G. Namer, *L'Abbé Leroy et ses amis: essai sur le Jansénisme extrémiste intramondain* (Paris, 1964). Highlights the variety of different strands within Jansenism during the classic period.

[45]   A. Sedgwick, *Jansenism in seventeenth-century France. Voices in the wilderness* (Charlottesville, Va., 1977). Solid and comprehensive, if rather bland, account of the heroic period. The most accessible account in English.

[46]   A. Sedgwick, *The Travails of Conscience: the Arnauld Family and the Ancien Régime* (Cambridge, Mass., 1998). Puts the crucial family of early Jansenism into context.

After *Unigenitus* in France

[47]   E. Appolis, 'L'histoire provinciale du Jansénisme au xviii siècle', *Annales, ESC* (1952).

[48]   D. A. Bell, *Lawyers and Citizens. The making of a political elite in old regime France* (New York, 1994). Excellent on the appropriation of Jansenist leadership by laymen.

[49]   P. R. Campbell, *Power and Politics in old regime France, 1720–1745* (London, 1996). Thoughtful analysis of the politics of a period dominated by the Jansenist problem.

[50]   D. A. Coward, 'The fortunes of a newspaper; the *Nouvelles Ecclésiastiques*, 1728–1803', *British Journal for Eighteenth-Century Studies*, 4 (1981), pp. 1–27.

[51]   J. Dedieu, 'L'agonie du Jansénisme (1715–1790)', *Revue d'Histoire de l'Eglise de France* (1928), pp. 161–214.

[52]   A. Farge, *Dire et mal dire. L'opinion publique au xviiie siècle* (Paris, 1992). Excellent on the appeal of the *Nouvelles Ecclésiastiques*.

[53]   Y. Fauchois, 'Jansénisme et politique au xviiie siècle; légitimation de l'Etat et délégitimation de la monarchie chez

G. N. Maultrot', *Revue d'histoire moderne et contemporaine*, 34 (1987), pp. 473–91.

[54]   D. Garrioch, *The Formation of the Parisian Bourgeoisie, 1690–1830* (Cambridge, Mass., 1996). Contains an important and readable section on lay involvement in Jansenism at parish level from the 1730s to the 1760s.

[55]   J. M. Grès-Gayer, 'The *Unigenitus* of Clement XI: a fresh look at the issues', *Theological Studies*, 49 (1988), pp. 259–82.

[56]   B. Groethuysen, *The Bourgeois. Catholicism versus capitalism in eighteenth-century France* (New York, 1962). Translation of a French work of 1927 which attempted somewhat crudely to fit changes in religious sentiment into social developments.

[57]   G. Hardy, *Le Cardinal de Fleury et le mouvement janséniste* (Paris, 1925).

[58]   D. C. Hudson, 'The Regent, Fleury, Jansenism and the Sorbonne', *French History*, 8 (1994), pp. 135–48.

[59]   B. R. Kreiser, *Miracles, Convulsions and Ecclesiastical Politics in eighteenth-century Paris* (Princeton, NJ, 1978). Superb analysis of the convulsions and their implications. Only superseded in detail by subsequent works.

[60]   J. McManners, 'Jansenism and politics in the eighteenth century', in D. Baker (ed.) *Church, Society and Politics (Studies in Church History, 12)* (Oxford, 1975), pp. 253–73. A sparkling and readable essay, subsequently expanded into over 200 pages of detail in volume 2 of *Church and Society in Eighteenth Century France* (Oxford, 1998).

[61]   C. L. Maire, *Les convulsionnaires de Saint-Médard. Miracles, convulsions et prophéties au xviiie siècle* (Paris, 1985). Reflective semi-documentary survey by the leading French scholar of a new generation.

[62]   C. L. Maire (ed.) *Jansénisme et Révolution (Chroniques de Port Royal, 39)* (Paris, 1990). The most important collection of essays on later Jansenism.

[63]   C. L. Maire, *De la cause de Dieu à la cause de la nation. Les Jansénistes au xviiie siècle* (Paris, 1998). The most authoritative and comprehensive treatment of later French Jansenism. Full of new insights and ideas.

[64]   J. W. Merrick, *The Desacralisation of the French Monarchy in the eighteenth century* (Baton Rouge, La., 1992). Important material backs a not always convincing argument.

[65]   M. J. Michel, 'Clergé et pastorale janséniste à Paris, 1669–1730', *Revue d'Histoire moderne et contemporaine* (1979). Indispensable statistical analysis of clerical support for Jansenism.

[66]   R. F. Necheles, *The Abbé Grégoire, 1787–1831. The odyssey of an egalitarian* (Westport, Conn., 1971).

[67]   R. R. Palmer, *Catholics and Unbelievers in eighteenth-century France* (Princeton, NJ, 1939). Puts later Jansenism in the context of the struggle against the Enlightenment.

[68]   E. Préclin, *Les Jansénistes du xviiie siècle et la constitution civile du clergé* (Paris, 1928). Classic analysis of links between Jansenism and the French Revolution. Now superseded at a number of points.

[69]   J. M. J. Rogister, *Louis XV and the Parlement of Paris, 1737–1755* (Cambridge, 1995). The day-to-day political background to the refusal of sacraments.

[70]   J. M. J. Rogister, 'A quest for peace in the Church: the Abbé A. J. C. Clément's journey to Rome of 1758', in N. Aston (ed.) *Religious Change in Europe, 1650–1914* (Oxford, 1997), pp. 103–33.

[71]   R. Shackleton, 'Jansenism and the Enlightenment', *Studies on Voltaire and the Eighteenth Century*, 87 (1967), pp. 1388–96.

[72]   J. Swann, *Politics and the Parlement of Paris under Louis XV, 1754–1774* (Cambridge, 1995). Takes in the story where [69] leaves it.

[73]   T. Tackett, *Priest and Parish in eighteenth-century France. A social and political study of the curés in a diocese of Dauphiné, 1750–1791* (Princeton, NJ, 1977).

[74]   T. Tackett, *Religion, Revolution and Regional Culture in eighteenth-century France. The Ecclesiastical Oath of 1791* (Princeton, NJ, 1985).

[75]   T. Tackett, *Becoming a Revolutionary. The deputies of the French National Assembly and the emergence of a revolutionary culture (1789–1790)* (Princeton, NJ, 1996). Includes the influence of Jansenists on the early revolution.

[76]   J. F. Thomas, *La querelle de l'Unigenitus* (Paris, 1950).

[77]   D. G. Thompson, 'The Lavalette Affair and the Jesuit Superiors', *French History*, 10 (1996), pp. 206–39.

[78]   D. K. Van Kley, *The Jansenists and the expulsion of the Jesuits from France, 1757–1765* (New Haven, Conn., 1975). A

fundamental reappraisal of the question which remains authoritative.

[79] D. K. Van Kley, *The Damiens Affair and the unraveling of the Ancien Regime, 1750–1770* (Princeton, NJ, 1984). Readable, stimulating, but not always convincing appraisal of the affair and its wider significance. The taproot of the idea of desacralisation of the monarchy.

[80] D. K. Van Kley, 'The Jansenist constitutional legacy in the French Revolution', in K. M. Baker (ed.) *The French Revolution and the Creation of Modern Political Culture, vol. I The Political Culture of the Old Regime* (Oxford, 1987), pp. 169–201.

[81] D. K. Van Kley, *The Religious Origins of the French Revolution. From Calvin to the Civil Constitution, 1560–1791* (New Haven, Conn., 1996). Largely recapitulates the author's previous arguments.

[82] M. Vovelle, *Piété baroque et déchristianisation en Provence au xviiie siècle* (Paris, 1973). An influential general argument about changes in religious sensibility, with important observations about the work of Jansenist bishops.

Jansenism outside France

[83] E. Appolis, *Entre Jansénistes et Zélanti. Le tiers parti catholique au xviiie siècle* (Paris, 1960). Important argument that not all those called Jansenists necessarily deserved the name, but had convergent though differently motivated priorities.

[84] E. Appolis, *Les Jansénistes espagnols* (Bordeaux, 1966). A pioneering study, but with conclusions liable to modification in the light of [95].

[85] D. E. D. Beales, *Joseph II, vol. I In the shadow of Maria Theresa (1741–1780)* (Cambridge, 1987). Chapters on Joseph's education and the meaning of 'Josephism' contain important insights on Jansenist influence in Austria.

[86] C. A. Bolton, *Church Reform in eighteenth-century Italy. The Synod of Pistoia, 1786* (The Hague, 1969). Full analysis of some key documents.

[87] W. J. Callahan and D. Higgs, *Church and Society in Catholic Europe of the eighteenth century* (Cambridge, 1979). Useful

collection of essays covering the background to Jansenism in this century.

[88]   O. Chadwick, *The Popes and European Revolution* (Oxford, 1981). Old-fashioned Anglican history of the best sort. A mine of heterogeneous information.

[89]   H. Gross, *Rome in the age of Enlightenment. The post-Tridentine syndrome and the ancien regime* (Cambridge, 1990), especially ch. 12.

[90]   A. C. Jemolo, *Il Giansenismo in Italia prima della Rivoluzione* (Bari, 1928). The starting point for all study of Italian Jansenism.

[91]   J. F. McMillan, 'Scottish Catholics and the Jansenist controversy: the case reopened', *The Innes Review*, 32 (1981), pp. 22–33. The first of three articles which offer valuable supplements to [4] on Jansenism in the English-speaking world. The later ones published in 1982 and 1988.

[92]   C. B. Moss, *The Old Catholic Movement* (London, 1948).

[93]   M. Nuttinck, *La vie et l'œuvre de Zeger-Bernard Van Espen. Un canoniste gallican et régalien à l'Université de Louvain (1646–1728)* (Paris, 1970).

[94]   L. J. Rogier and P. Brachin, *Histoire du Catholicisme hollandais depuis le xvi siècle* (Paris, 1974).

[95]   J. Saugnieux, *Le Jansénisme espagnol du xviiie siècle: ses composantes et ses sources* (Oviedo, 1976).

[96]   M. Vaussard, *Jansénisme et Gallicanisme. Aux origines religieuses du Risorgimento* (Paris, 1959).

Index

Bulls – *continued*
   In Eminenti (1643), 24, 26
   Provisionis Nostrae (1579), 9
   Regiminis Apostolici (1665), 33, 43
   Unigenitus (1713), 1, 2, 3, 4, 45–58,
      60, 61, 64, 66, 69, 70, 75, 76,
      80, 85, 88
   Vineam Domini (1705), 43, 49

cahiers of 1789, 81, 83
Calvin, Jean, 6, 20, 22, 48
Calvinism, 1, 6, 11, 17, 22, 24, 35,
      38, 80, 87
Cambrai, 42
Canterbury, 80
Carré de Montgéron, Louis-Basile,
      56, 60, 61
Carthage, 5
casuistry, 13–14, 17, 29, 30, 36
Champaigne, Philippe de, 2
Channel, English, 80
Charles III, 68, 74
Chauvelin, Henri-Philippe, 72
Chinese rites, 42, 49, 71
Choiseul, duke de, 66, 71
Christ, 5, 6, 15
Cistercians, 16, 46
civil constitution of the clergy,
      3, 83–4
clergy, parish, 7, 26, 30, 36, 37,
      47, 48, 54, 60, 63, 77, 78, 87
clerical oath (1791), 84
Codde, Pieter, 70
Coffin, Charles, 62
Colbert, Charles-Joachim, 59
concordat (1801), 86
Conti, prince de, 64
contrition, 19, 36, 45
convulsionism, 2, 3, 56–8

Damiens, Robert François, 65–6
Dauphiné, 81
De auxiliis, congregation, 11, 24
dechristianisation, 85
desacralisation, 4, 66, 88–9
despotism, 62, 80, 81, 82, 83
dévots, 15, 16, 19, 20, 62,
      64, 66, 73, 79

Dominicans, 8, 10, 11, 36,
      47, 48, 49, 68
Dort, Synod of, 11, 17
Duguet, Jean-Joseph, 41,
      48, 49, 50, 81–2
Dutch republic, 41, 44, 69–70

Eden, Garden of, 5
Embrun, 53
Ems, *Punctation* of, 75
enlightenment, 4, 56, 73, 74, 88, 89
Erastianism, 69, 75, 85
Escobar, Antonio, 29
estates-general, 81, 83–4
Etemare, Jean-Baptiste, 49, 70
Ex Omnibus (1756), 64, 65

Febronius, Febronianism, 69,
      74–5, 79, 80, 88
Fénelon, François de Salignac
      de la Mothe, 42, 44
feuillle des bénéfices, 59
figurism, 49, 50, 55–6, 70
five propositions, the, 25–8,
      32, 38, 42, 45
Fleury, André Hercule, cardinal de,
      52, 53, 54, 56, 57, 59, 60, 61, 62,
      63
Florence, 76, 77
formulary, 27, 31–4, 37,
      39, 41, 43, 83, 84
   of Alexander VII, 33, 37, 42, 52
France, 1, 2, 12–14, 24, 31, 37,
      40, 41, 43, 44, 47, 67, 68, 71, 72,
      73, 74, 76, 79, 80, 85, 88
Francis II, 85
François de Salles, St, 19
freemasons, 83
free will, 1, 8, 10, 88
French Revolution, 2, 4, 64,
      79, 83–6, 88, 89
Fronde, 25, 26, 48

Gallican Articles of 1682, 40, 46, 85
gallicanism, 14, 23, 27, 31, 37, 40,
      46, 50, 51, 52, 67–8, 75, 85–6, 88
Gazier, Augustin, 2, 3
general hospital, 62–3

Germany, 69, 74–5, 88
God, 5, 6, 8, 10, 19, 21, 35, 55
Goldmann, Lucien, 3, 89
Gondi, Jean-François de, 25
grace, 6, 7, 8, 9, 11, 19, 21, 29, 45
grand remonstrances, 63, 66
grand siècle, 2
Grégoire, Henri, 80, 83, 84, 86
guichet, journée du, 16

Habert, Isaac, 26
Habsburgs, 9, 20, 74–5
Henry III, 12
Henry IV, 12–13, 14, 15
Hippo, 5
Holy Roman Empire, 12, 75, 77
Holy Sacrament, Institute of, 16
Holy Thorn, miracle of, 29, 31, 55
Hontheim, Nikolaus von, *see*
    Febronius
Huguenots, 40, 80

iconoclasm, 85
Index, 7, 39
Inquisition, 11, 24, 27, 31, 44, 68
Italy, 12, 74, 75–8, 79, 85, 88

Jacobins, 85
Jansen, Cornelius, 1, 17–18, 20,
    21–2, 23, 24, 25, 27, 28, 32, 33,
    35, 36, 45, 47, 48, 68, 87
Jesuits, 1, 4, 8, 9, 10, 13, 14, 15, 16,
    17, 18, 19, 21, 22, 23, 24, 26, 29,
    30, 32, 33, 35, 37, 39, 42, 43, 44,
    48–9, 65, 68, 69, 70, 71–4, 76,
    77, 78, 79, 82, 86, 87, 88
Jews, 80
Joseph, Father, 19–20
Joseph I, 71
Joseph II, 69, 75, 76, 88

La Borde, Vivien de, 49
Langres, 16
Latin, 7, 18, 23, 29, 30, 38
Lavalette, Antoine de, 72
Laxism, 30, 36
League, Catholic, 12–13, 15
Le Gros, Nicolas, 50

Le Maître, Antoine, 19, 24
Leopold II, 76, 77, 85
Le Paige, Louis-Adrien, 63–4,
    66, 73, 80, 82, 84, 86
Lessius, 10, 11, 17
Le Tellier, Michel, 44, 50
Liancourt, duke de, 27
Lisbon, 71
little schools, 24, 31, 33, 37
Longueville, duchess de, 38, 40
Lorraine, 80
Louis XIII, 14, 25, 81
Louis XIV, 30, 32, 33, 38, 39, 40, 41,
    42, 43, 44, 45, 46, 48, 49, 51, 54,
    68
Louis XV, 50, 63, 65, 82, 89
Louis XVI, 82, 83
Louvain, 9–11, 17, 18, 21, 22, 23, 41,
    68–9, 74
Loyola, Ignatius, 8, 13
Luçon, 19
Luther, Martin, 6, 7, 8, 9, 20

Madrid, 18
Maire, Cathérine, 3
Mariana, Juan, 13
Maria Theresia, 69, 74
Marseille, 72
Mars Gallicus, 18, 20, 22
Martinique, 72
Marxism, 3, 89
Maultrot, Gabriel Nicolas,
    66, 80, 81, 84
Maupeou, René Nicolas Charles
    Augustin de, 79, 82
Mazarin, cardinal Jules, 23, 25, 26,
    27, 30, 31, 32, 46, 55
Mey, Claude, 66
Milan, 23
Mirepoix, 61
Molina, Luis de, 10, 11
molinism, 10–11, 73, 88
Montpellier, 59
Munich, 75
Muratori, Lodovico Antonio, 76

Netherlands, 11, 39, 41, 42, 68
New Testament, 39, 47